T0091600

Making Sense of the Senses

Making Sense of the Senses

Tobias Wibble

Karolinska Institutet, Sweden

 **World Scientific**

NEW JERSEY · LONDON · SINGAPORE · BEIJING · SHANGHAI · HONG KONG · TAIPEI · CHENNAI · TOKYO

Published by

World Scientific Publishing Co. Pte. Ltd.

5 Toh Tuck Link, Singapore 596224

USA office: 27 Warren Street, Suite 401-402, Hackensack, NJ 07601

UK office: 57 Shelton Street, Covent Garden, London WC2H 9HE

British Library Cataloguing-in-Publication Data
A catalogue record for this book is available from the British Library.

MAKING SENSE OF THE SENSES

ISBN 978-981-124-629-6 (hardcover)
ISBN 978-981-124-630-2 (ebook for institutions)
ISBN 978-981-124-631-9 (ebook for individuals)

For any available supplementary material, please visit
https://www.worldscientific.com/worldscibooks/10.1142/12527#t=suppl

Typeset by Stallion Press
Email: enquiries@stallionpress.com

To my Ardita, who makes everything make sense

Acknowledgement

It is more than fair to say that without my partner Ardita this book would never have happened, and I cannot thank you enough for the countless hours spent reading through every draft I pulled together and for staying by my side throughout this adventure. I would also like to thank my grandparents Gudrun and Ernst for instilling in me the importance of learning and joy of storytelling, and my parents Tomas and Marita. A great portion of this book was written in London and I would like to thank Sabri and Dylbere for welcoming me into your family, and Anisa and Anilda for always being there just down the corridor, with a special mention to Lyra for being a very good dog.

I would also like to thank everyone at Karolinska Institutet who has contributed towards my understanding of sensory neuroscience. While this continues to be an ongoing effort, it would not have been made possible without the support of Prof. Sten Grillner, whom I will never be able to thank enough for giving me the chance to try my hand at science as a profession. I would also not have gone through this journey of cross-species physiology had it not been for the support and friendship of Prof. Juan Pérez-Fernández, who continues to expand the boundaries of human knowledge presented in this book. This book would also not have been possible without the backing of Prof. Tony Pansell and the team at Marianne Bernadotte

Centrum, with a special mention to Prof. Jan Ygge and Prof. Gunnar Lennerstrand for their continuous support. I would also like to thank World Scientific for their time in making this text into an actual book, and especially Ms. Joy Quek for her clear and structured feedback throughout this entire process. Finally, I would like to extend a special thank you to Prof. David Zee for lending his ear, and eyes, to the manuscript that became this book; it is difficult to find a higher merited scholar on the topic of sensory neuroscience, and I am very grateful that this book may start with such a distinguished preface.

Foreword

Tobias Wibble writes "If you are in high school or in your first years at university and interested in biology, this book is for you." After reading the book, I would rewrite this sentence, " If you are curious about any of the following; the biology of the human body and especially the brain and its peripheral appendages, the remarkable way animals have evolved to hold their place on the planet, and what the future holds for how human beings can use science to ensure our future place in the environment, this book is for you". Even as a neurologist, long interested in the human brain, and especially how evolution has influenced both normal and abnormal brain function, I learned much from this book about the fascinating ways that nature has shaped the bodies and brains of all living creatures for survival and reproduction. This book emphasizes how we acquire information about the environment though our sensory appendages, our ears, eyes, nose and mouth, skin and joints, and then compares our often relatively feeble capabilities with some of the champions of sensation in the animal kingdom. Moths are the best listeners in the world. They can discern frequencies more than ten times higher than we can, making it possible for them to detect the echolocations of predator bats. Dolphins can emit vibrations in water and then detect echoes that allows them to do the equivalent of a noninvasive ultrasound evaluation or CT scan of a human's internal organs (since we are mostly made of water). And we learn of more surprising capabilities of animals in vision, smell, taste, and touch. For those interested in the history of science, this book also gives some background on how the ancient philosophers developed their ideas of how we perceive our world. For those interested in the future of science, there are sections on how our sensory capabilities can be improved or restored with prostheses, bionics, and robotics. Like any engaging lecture to a diverse audience, there is something in this book to take home for everybody, no matter what their previous level of scientific knowledge or general intellectual interests.

David S. Zee, MD

David Zee MD and Paul and Betty Cinquegrana Professor
Departments of Neurology, Neuroscience, Ophthalmology,
Otolaryngology-Head and Neck Surgery
Division of Neuro-Visual and Vestibular Disorders
The Johns Hopkins University School of Medicine

Contents

Introduction

I am of the conviction that a story of science, if well told, will leave the reader in a confusing state of wonder combined with just a hint of existential insignificance. I have no ambition of rocking the foundation of your relationship with Nature, but I do hope to at least nudge the pedestal that mankind has placed itself on. My much simpler primary goal however is to present some much needed context on the way we perceive the world through our senses in the form of sight, hearing, smell, taste, touch, and more.

Clearly, we have done this rather well, or we would not find ourselves the most invasive species to ever take shape; not even the seemingly indestructible cockroach can muster the determination to build and inhabit metal cubicles circling Earth through space at 8 km/s. Despite this, our human tools of sensory information collection are by no means the pinnacle of evolution, at least not on any individual merit. This sensory inferiority does not end there. Several animals possess senses that took us eons of evolution to even conceptualise, and it seems quite likely there are phenomena out there we still are not fully aware of. Sure, we can record infrared or use radar, but even then all of our gadgets are designed to serve our senses such as vision, touch or hearing; interestingly, very seldom are taste or smell incorporated in our manufactured tools, though there are some notable exceptions which we will go into later.

The layout of this book is quite straightforward. First, we will go through each of the human senses, and to avoid accusations of favouring one system over the other, each sense will be dedicated a chapter. First, we will explore how we ended up with that particular sense to start with. For example, what brought on the need for smell? Surely, not being able to smell the sulphuric pits of young Earth would have been preferable? While the first mutation that gave rise to any particular new sense might have been due to chance, evolution took these sentences written in the sand and sculpted them into the granite of our DNA. As you might imagine, creating entire new sensory systems is a costly affair and they all exist for a reason. After evolution has had its turn, we will go into the broad neural mechanisms that make each sense work, for instance shedding some light onto how particles emitted by the sun are translated into electric impulses for our brains to make sense of. After having gone through the five major senses we will move on to some additional ways through which we humans perceive the world.

Having followed the pathway of sensory input to perception as allowed by our human physiology, we will explore how other animals use each particular sense. Humans are relatively decent at collecting and interpreting sensory data, but we are simply mediocre compared to many of our animal cousins. Biological solutions for perceiving the world extend far beyond what we humans can see, hear, smell, taste or touch and indeed, writing a book on the senses and only focusing on the very limited human experience would be myopic. In addition to these five traditional senses, we will therefore be spending some time exploring the more esoteric sensory systems of the animal kingdom. We will attempt to figure out how dolphins possess X-ray vision, birds use their beaks as magnetic compasses, and spiders fly on electrical currents.

Throughout this book, we explore the physiological make-up of each sensory system. It is my ambition to provide references for all claims, both to ascertain that the facts can be trusted, but also to allow people the opportunity to dig deeper into a specific topic.

We will see plenty of references from *Principles of Neural Science*[1] by Kandel, Jessel and Schwartz. Its impressive size reflects its breadth, covering all aspects of neuroscience and certainly the general mechanisms that we will discuss in these chapters, and the first thing I did after getting a job in neuroscience was to run to the university bookstore and get my own copy of this encompassing work. So instead of adding a reference to it after every sentence, I will introduce it now and retouch on it within each chapter when introducing the basic mechanisms of each sensory system, because it deserves all the credit it can get.

At this point, I feel it is important to announce what type of reader this book has in mind. In other words, who do I imagine you are, sitting down reading this sentence? If you are in high school or in your first years at university and interested in biology, this book is for you. Though I do hope that even if you left your student days many years ago and are simply curious about the world around you, that you may found this sensory odyssey across the animal kingdom interesting. We will cover a lot of ground in fairly few pages and as this is an introduction to the field, there might be a few oversimplifications. This book is therefore geared towards the general reader rather than experts in the field, and with any luck it might inspire someone to delve deeper into sensory neuroscience and evolution. I can only hope that some of the wonderment I felt when researching these chapters shines through the pages. With that being said, let us get started setting up the framework for our forthcoming journey.

You cannot really talk about the senses in any comprehensible way without establishing what constitutes a sense in the first place. Most of us grow up learning that humans possess five main senses, but at the very least, this is a heavily biased cultural interpretation of biology, influenced by various religious dogmas as well as personal interference from some of the great thinkers of history. Broadly speaking, a sense is characterised as an organism's ability to physiologically gather information that influences its perception of the world, giving us the opportunity to understand or react to it.[2] Take a millipede making its way across a branch. We can see its little legs working, hear their gentle tapping on the wood, feel its contours on the tip of our finger and smell its distinctly bad odour, signalling that tasting it might be a bad idea. Should we ignore this warning and put it in our mouth, our sense of taste would tell us that it possesses a crunchy and bitter exoskeleton. Had this particular millipede emigrated from Chernobyl, we would never have known the levels of radioactivity emitting from it, as we lack the appropriate sense to detect gamma radiation. Our perception of the world is therefore limited by our physiological capacity to interpret it.

So are we limited in our biological perception by these five essential data-collection systems? This is certainly not the case. As we shall see, there are several contenders for other mechanisms we may call senses, such as perception of balance which has a whole organ dedicated to it. As it turns out, humans have several senses that we never really learn about in school. This book aims to remedy this by also exploring these other physiological methods of perception which are less talked about. One important question-mark might be floating in your head right now, as it did for me when I first set out to write these words: who decided that we have five senses? Where does that classification come from, and at what point in history did we agree that "actually, five senses seems reasonable, let's not push it"? I think the answer to this might be somewhat of a surprise to many. Let us therefore start there, and try to make sense of how it all began.

Chapter 1

Making the Senses

Humans have always possessed the finely tuned arts of vision, hearing, smell, taste and touch. Any person questioning that may as well question the shape of our planet. So how could there possibly have been any changes over time for what we consider to be the basic human senses? Much like it took us some time to figure out that the Earth is round, we needed some thinking time before we could ponder over the nature of our tools of perception. As we move through subsequent chapters, we will explore how each of these senses biologically evolved, but for now let us instead delve into how we humans made the very idea of the sensory systems come to life in the first place.

As with most things in science history, we find ourselves going back to the ancient Greeks. Written during this period was a manuscript called "Sense and Sensibilia",[3] a work that was given homage much later in the title of a classic Jane Austen novel. This text dealt with human sensory perception in a rather philosophical way. The author of this work was none other than Aristotle himself. Instead of providing us with the current Five, he instead introduced four. In true ancient spirit, these were all matched with one of the basic elements: sight was water due to the gushy contents of the eye, sound was air, smell was fire, and touch represented earth. So what about taste? Why was this modern representative of the senses treated like an estranged cousin? Taste, Aristotle argued, is just another aspect of touch. Today, this might seem somewhat strange. Taste is such an essential aspect of life, and probably one of the most lucrative senses to exploit in true entrepreneurial spirit, though in all earnest, taste is not the same as flavour but that is for a later chapter. Before we judge Aristotle, let us first view ourselves through the looking glass instead. Maybe the same question of why some senses are not recognised will be asked in another 2,000 years, where they may also wonder why we did not allow temperature the status of Sense. Our ability to tell hot from cold has a completely different physiological mechanism than the pressure-aspect of touch. Nevertheless, we use "touch" as a sort of basket where we keep our uncategorized senses all jumbled together. We will go further into this in a later chapter, but for now let us at least cut Aristotle some slack.

Aristotle was actually quite a bit more generous than his spiritual predecessor, Democritus, who had set the precedent on what should be considered a proper sense.[2] In Democritus' time, which was one generation earlier to Aristotle's, he argued for a world without sight, hearing or smell. As a philosopher, he clearly possessed an amazing mind and was the first to suggest the shape of atoms. However, he had very little faith in the sensory systems, calling them the "bastard" senses,[4] fallible due to human subjectivity compared to the pure truths he was able to conjure in his mind. Democritus did not argue that people cannot see, hear or smell. He postulated that we possessed senses which we lacked the internal capacity to process in any meaningful and objective way, thus concluding that only touch remains a proper method of perceiving the world.

It would be another 500 years before the foundation of the Chosen Five that we today consider to be innately obvious. Aulus Gellius was a Roman author living in the middle of the 2nd century who mostly focused on grammar and antiquities.[2] He importantly preserved several works of his contemporaries, but also shared his own philosophical thoughts on certain matters. For one, he claimed that there are only two colours, *rufus (red)* and *viridis (green-yellow-blueish)*.[2] While this obviously failed to catch on, his identification of five distinct senses would influence the way we look at the world even today, 2,000 years later. Gellius' addition of taste onto the Aristotelian senses might not have been as much a silent revolt against his Greek predecessor as him simply doing what he loved most: preserving the thoughts of his time. Either way, it is clear that it was around this time that Western culture became acquainted with the Big Five of perception. I say *Western* because there are much, much earlier depictions of the senses in the Hindu Katha Upanishad, written in the 6th century BC.[5] Being fans of allegory, the Indian manuscript depicts a chariot, representing the mind, being pulled by five horses corresponding to our biological tools of perception. This is quite an apt simile — the mind only being able to move thanks to the methods through which it gathers information about the world.

It seems unlikely that these two converging views, Western and Eastern, should have influenced one another. Distilled down to the essentials however, these definitions of what a sense was were fairly similar despite originating in disparate parts of the world, highlighting a universality to how we perceived our own perception. Much like in India, the Western school of thought would come to involve religious themes. Gellius was objective enough to credit Mother Nature with our sensory gifts, but as we move towards the Middle Ages, we need to involve the Catholic Church. It was primarily church officials who had the privilege of having the time to do any real philosophizing, and the only ones who could write it down. So despite the oftentimes strained relationship between religion and science, we find most of the important academic findings to be somehow related to The Vatican's Holy Seal.

It would appear that the senses make intermittent appearances throughout Christian services as they spread throughout Europe.[6] Again, this most likely was not any attempt to cement what the senses should be, but rather a reflection of general public opinion. By connecting divine attributes to the senses, it arguably provided the Church with an approachable way of communicating their message. In the 11th century, the monk, and later saint, Peter Damian attempted to further humanity's connection with God by claiming that the five wounds suffered by Jesus on the cross were inflicted as a reference to the senses.[2] Considering that Damian resides in the highest level of paradise, at least according to Dante's Divine Comedy, it seems that our sensory classification has indeed accrued some divine interference. A spiritual successor, the English friar John Bromyard, explicitly adopts this line of thought during the 14th century.[7] At roughly the same time, and perhaps most importantly, one of the most read medieval authors, the mystic Richard Rolle, echoed this link between humanity and the divine.[2]

The frequent presence of theologians in the history of the senses is of some importance — people did not go into conflict with the Catholic Church unless they had some pretty strong convictions. Still, we do not yet necessarily have the term "senses" in place. An ambitious project to put world history in writing, told from a deeply religious perspective and published by an unknown author in the 14th century, tells us more about how the senses were perceived. In the *Cursor Mundi*,[8] five outward "wits" are described: *hering, sight, smelling, fele,* and *cheuing.* You might be able to guess which one corresponds to which modern version of each sense.

A quizzical reader might now wonder: if there were "outward" wits, does that mean there were also "inward" senses? Here we find a pivotal moment in our story. Thus far, we have only dealt with the outward wits because they were the ones which came to be adopted as proper senses. During Aristotle's time, there already existed philosophical internal counterparts in the form of "internal wits".[9] These were faculties of the soul, representing the pure logical capacity of the human mind that was so highly regarded by the ancient scholars. Geoffrey Chaucer himself had translated a Roman source on these five introspective senses, labelling them as : *comoun witt, ymaginacioun, ffantasye, estimacious,* and *mynde.*[10] In this case, I will take the liberty to provide the translations: common wit, imagination, fantasy, estimation and memory.

One of these particularly stands out. The "common witt" sounds awfully familiar to "common sense", and they likely refer to the same thing. What prompted the transition from witt to sense? It would seem that the two words were synonymous in Middle English.[11] With time, this shared meaning drifted apart, and sometime during the 17th century it would seem that the term "sense" now referred to the outward wits, while "witt" itself dealt with the internal faculties.[11] The phrase "common sense" apparently survived this transformation and clearly hints at its own history to this day. From the other perspective, we can see how calling someone "witty" might be one of few contemporary references in our daily speech that alludes back to more complex philosophical times.

There is further evidence that the introvert methods of perception were referenced together with the outer senses even earlier. Plato, echoing the words of his mentor Socrates, wrote of the "innumerable senses" giving examples like *pleasure* and *fear* together with the traditional *sight, sense* and *smell.*[2] So why was it that this deeper meaning of the sensory concept was abandoned? One can speculate, but one of the most likely reasons could be that it never caught on with the general public. The Outer Five are far more tangible — you can see a person's nose, ears, eyes, mouth and skin, all correlating to one specific mode of perception. Discussing the inner systems of deduction with a starving peasant who just lost his entire family to the plague would be both futile and inconsiderate. Most, if any, education would have been through the Catholic Church which did not necessarily want the peasantry to think too hard about their bodily functions. "Jesus had five wounds and so you have five senses, so don't try to get witty with all these wits" may have been more than enough in terms of sensory neuroscience education.

The outward senses also make much better topics in popular culture, with popular culture in this case referring to the contemporary Baroque paintings of the 17th century, coinciding well with the separation of wit and sense.[12] The allegorical depictions of their outward forms were well suited for conveying a deeper message — mirrors, clear streams of water, and of course eyes, were clear motifs suitable for the somewhat extravagant style of the Baroque paintings. By comparison, working with the inner senses through a visual medium is somewhat of a challenge. If you ever play charades with a group of friends, I dare you to try acting out "imagination".

From there on, the cultural avalanche gained momentum, sweeping away the very notion that humans could possess something as foreign as a "sixth sense". There was no malice in writing out these additional senses from history and mainstream culture. There was never any Machiavellian king or scholar that one day decided to condemn all the followers of thermoception to the gallows. Things simply developed on their own. Just as evolution has provided us with our senses, thousands of years of human civilization have produced a set of easily identifiable references to them. With that being said, let us leave the door ajar for new ideas and interpretations.

In essence, our senses are as much culture as they are biology. While the lenses provided by these fields differ vastly in their time-frame, both are subject to change, with change being integral to both of them. Let us stop philosophizing for a bit and explore the neural systems that allow our brains to interpret the outside world. As we shall see, they are far from the perfect infallible machines we often mistake them for. So let us take a journey, starting from the slimy moulds inhabiting our world while it was still young. We will explore the intricate mechanisms as well as the peculiarities of the highly specialized lifeforms with which we are lucky enough to share this space and time. Finally, we will look into the almost supernatural abilities of our biological contemporaries, who through evolutionary prowess have mastered arts only available to us clever apes through modern computers.

Let us begin.

Chapter 2

Vision

Let There be Light

The eyes are often depicted as the epitome of evolution, the prime example of organic perfection, a window into the soul. For some, this is evidence enough for an intelligent designer, because how could something so complex happen through accident? Let us walk through why, and more importantly how, our eyes decided to emerge out of the relative safety of the head.

Firstly, our journey takes us back to the primordial soup, or rather, just after it has cooled. Organic life has taken to the planet as fish to water, quite literally. Light flickers on the endless oceans, circling the one great landmass that over the millennia will spread to make up our current continents. But for now, under the sea, the light shines on a small fish-like creature, not much bigger than a fingernail. It floats gently through the water, almost grazing the rocky bottom a couple of metres below the surface. Every now and then the creature stops, getting a feel for the vibrations in the waves, ready to dash for its life if needed. Albeit small, a gentle shadow moves across the ocean floor. The blind little fingernail fish does not see the suspicious outcrop underneath. As the shadow moves across it, there is a small movement in the sand. Our fish notices the movement too late as the suspicious outcrop takes shape. Soon, where there were two animals there is now only one, and for that, one of them can thank its primordial form of vision. A new sensory organ has come to life.

The example above is simply a dramatisation of how vision could have given our predecessors a clear advantage on the evolutionary ladder. We cannot really be sure of what animal was first gifted with the ability to see but it seems to have happened roughly 700 million years ago, before the split of Bilateria (animals with bilateral symmetry) and Cnidaria (such as jellyfish).[13] This can be compared to the age of modern eyes, of which the earliest evidence is from a 530 million-year-old fossil of the now-extinct trilobite.[14] As the sun has been there from the very start, it is not surprising that the need to notice its effect was needed and, through evolution, was acquired by Earth's inhabitants. While the sea creature above, the one masquerading as a suspicious-looking outcrop, perhaps lacked conventional eyes, it could still sense changes in light with the help of proteins that were highly specialised to detect photons, the latter being the mixture of particle and wave emitted by the sun that makes up all forms of light. This forms the basis of our eyes — a thin sheet of cells detecting contrasts between light and dark. Eventually, these light-detecting proteins will form our retina, but that is still some million years in the future.

Some may think it strange that proteins can detect light. Obviously there are some fundamental differences between proteins in our food and those that form structures in our eyes — proteins as a food source are usually in the shape of muscle fibres, while the photoreceptive, or light-detecting, proteins would make up a rather meagre meal. Essentially, the function of these proteins is determined by their genetic composition or DNA. Luckily for us, DNA changes over time and mutations, or changes, to the DNA are the basis for evolution and the reason why our eyes are not that original thin sheet of proteins only capable of detecting shadows.

A Window Takes Shape

The basis of evolution lies in a very simple truth: Genetic success can be measured in how many kids you have that live long enough to have offspring of their own. So a mutation leading to something enabling you to have more offspring will, in the long run, be part of the evolutionary pathway for that creature. The same is true for eyes. While the simple eye in the previous example definitely helped, it would not have been very good at anything other than letting its owner know if the region above it was dark or light. The next step, albeit small, would rectify that. Generations and several DNA mutations down the line, a small indentation, just beyond the edges of the early retina, created a crater in which the light-sensitive proteins rested.[13] This gentle change opened up a whole new world of possibilities. As the light now entered the crater at different angles, the perception of the world came with a sense of direction. If light hit the bottom left of the crater, it meant that the light must have come from the top right! The image would of course be upside down, but that is nothing the brain could not handle by simply flipping it.

With this new visual advantage, the creature manages to survive and soon the seas are teeming with wildlife that can see. Over the generations, the gentle indentation will grow, or rather sink deeper, into the head. Generations down the line this crater has almost become a cave, with nothing but a small opening at the top. The small mouth of this cave allows the brain to infer the direction of any incoming light with a very high degree of accuracy due to the inclination of the light reaching the retina. Instead of just letting the fish know what is going on straight above it, these new eyes will tell a story of *where* something is and even *what* it is. The latter is based on the outline of objects in the visual field; as a big crater with a small opening allows for computing angles more precisely, the brain can simply use this information, do some quick architectural math, and build a visualised model.

So now the eye can detect where, and to a certain extent what, an object is. The vision is still going to be somewhat blurry. The light, with its beams crisscrossing through the water after having been refracted on the surface, will enter the eye unfocused. This is where the crater becomes a cave, with a little glass ceiling where the opening used to be. This glass ceiling, which really is just a see-through continuation of the eye itself, is the first lens. As the descendants of our seeing ancestors make their way out of the oceans and onto land, this lens, and the eye, will continue to evolve. While it might seem complex, the basic concept of how these organic gadgets work is essentially the same as that of a magnifying glass. Hold one up to the sun and the light will converge on a point on the other side. Depending on the distance to this object, the focus of the light can either be very broad or very narrow, and if you are not careful, you will soon see smoke rising from whatever you are pointing the light at! The lens of the eye works in a similar way, refracting light in order to focus it to a specific point at the back of the eye. As this point is very narrow, the eye is allowed to focus its main function on this limited region of the retina. Keeping an object focused on this point is what keeps our human vision from being blurred and allows us to focus on comprehensible units of visual information rather than being bombarded with light from all sides without being able to select where our main focus should lie. This is what gives us an excellent capacity for sensing edges and contrast. As we shall soon discover, the answer to why this is lies in the type of cells we find here.

We will leave fish aside for now and fast-forward through the millennia. Depending on the creature and the lens, the eye will come to look very different between species in a process called *convergent evolution*. This means the evolutionary road towards eyes and vision is not unique.[13] There are in fact many roads leading to the same place, though they may look vastly different. We will discuss how different animals use their eyes in a later chapter, but firstly, let us focus on our own, human eyes.

Through the Looking Glass

We have established how our eyes came to be: a thin sheet of light-sensitive proteins slowly sank into the head, eventually being encompassed by it, leading to what we would come to call an eye. While this allowed us to perceive what and where something was happening, what did this mean for colours? And how was the light that hit the retina converted into an electrical signal for our nerves to send to the brain? And once it received the signal, how did the brain know what to do with it?

Let us start with colours. The photoreceptors, or the light-sensitive pieces of protein, lie at the back of the eye.[1] Light, however, is not a singular entity, but behaves both as a set of particles as well as a wave. We will not go into the physics of light, but suffice to say that the wave format of light exists in different wavelengths, some short and others longer. This is the basis of colour: light with a short wavelength we call blue, and light with a long wavelength we call red. All objects interact with light in some way. A tomato reflects light of long wavelengths, which means the light escaping it is red. Similarly, if an object is black it means that it has not reflected any light, while a white object is more generous with all wavelengths being bounced back into the world.

So how do our eyes distinguish between these wavelengths? Essentially, there are two types of photoreceptors: rods and cones. *The rods* are the more numerous of the two, with 90 million rods inhabiting the retina. They cannot, however, tell the difference between colours. The rods are like the early eye of our fishy ancestors, being able to tell if there is a light shining on them or not. The privilege of being able to enjoy the colourful things in life lies strictly with the other group of photoreceptors, *the cones*.

The cones could be seen as the upper class of the photoreceptive world. They make up about 5% of the population with their 4.5 million representatives, and more importantly they primarily reside in the privileged, well-lit, part of town. Although cones can be found across the retina, they are concentrated in the central fovea, the area of the retina on which the lens focuses light onto. Reflecting the nature of this location, the cones express certain properties.

While both rods and cones sense light through tiny membranous disks, it is due to differences in these that the two photoreceptors vary. The disks making up the rods and cones contain retinal, derived from vitamin A. In case you wondered why carrots are supposed to be good for your eyes, this is why, as vitamin A is absolutely necessary for the photoreceptors to work. Interestingly, we know that *placozoa*, one of the simplest forms of life discovered, had light-sensitive proteins[15] but it lacked the capacity of incorporating retinal, the vitamin A derivative, into the disks. We cannot be quite sure how useful these types of proto-eyes would be to the placozoa. Nevertheless, we should be thankful for whatever use it made out of it, as it provided the foundation for our own vision to take shape.

The nature of the disks distinguishes rods and cones from each other. While the rods are excellent at detecting contrasts between light and dark, the disks of the cones can detect colour.[1] There are three types of cones that sense different colours, divided into those that can detect short, medium, and long wavelengths. Variations in their disks is what makes them sensitive to different wavelengths of light. The disks on both rods and cones only have a life span of 12 days however, meaning that as new disks are formed on the base of the photoreceptor, those at the tip need to fall off like leaves from a tree. These are absorbed by the pigmented cell layer that gives the retina its darker colour.

This means that cones have a special predilection for blue, green or red, activating in response to the different wavelengths of light that correspond to these specific colours. Consequently, we humans possess what is called trichromatic colour vision, as we can see three different colours. All other colours, like purple or orange, are just combinations from multiple cones trying to convince the brain that their signal is the right one. The brain, being the

diplomat it is, adjusts these signals so as to make everyone somewhat pleased. For example, when both the red and blue cones
activate, the brain will combine these signals to make you see purple. You have probably heard of other types of light, such as
ultraviolet (UV), infrared (IR), or gamma radiation. Technically,
for a wavelength to be called *light* it needs to be on the visual spectrum for us humans. This anthropocentric perspective means that
there is no such thing as UV or IR light, only radiation. Instead,
these wavelengths follow the same laws as the light we see, but due
to our evolutionary beginnings we cannot see them. One possible
explanation for this I find rather compelling: the light we can
see — blue, red, green and their combinations — make up a chosen few wavelengths that are refracted in water rather than being
absorbed by it.[16] Given our eyes first evolved in water, this seems to
be too much of a coincidence.

So how are those beams of sunlight translated into neural
impulses that allow us to see an object? I will outline the basic principles here, but it may be helpful to take a look to Figure 1 to have
it visualised. When beams of light hit something, say a red apple,
light is reflected so only long wavelengths will bounce off the surface
and hit our eyes. The waves first hit the translucent part of the eye,
the cornea, which we can liken to the roof of the great cave that is
the eye bulb.[17] This is the first step in which the light is refracted,
similar to the example with the magnifying glass, as it enters a liquid-
filled chamber in a more concentrated beam. This is the first of two
chambers that make up the biggest part of the eye. Having passed
through it, light makes contact with the lens and its path is further
adjusted so that it will hit the region where the colour-sensitive
cones are concentrated, the fovea. The lens makes sure this happens
by changing its shape. Tiny muscles, part of what is called the ciliary
body, tug at the lens making it either thinner or thicker depending
on how much the incoming light needs to be refracted, with the
photoreceptors in the retina providing some feedback on how much
adjustment is needed.

Going back to the apple, the shape the lens needs to take depends primarily on how far away the apple is. At baseline the lens has a rather round shape, almost like a marble If the apple is close, the tiny muscles will get a workout and contract, making the lens thin. An object far away would instead require a thicker lens and the muscles get to rest. Therefore, every time you focus on something, your eyes quite literally get a workout. Do this for long periods of time and the muscles will start to cramp. Eventually, the muscles will have a harder time adjusting the lens, and as a result, the light will not be able to hit the right spot at the back of the eye, which is why vision gets less sharp with age; clearly, this leaves some room for improvement if we were to redesign our own methods of seeing.

The light beam then passes through the second, bigger chamber. The fluid, like that found in the previous chamber, is called *aqueous humour* and flows freely between the chambers through the pupil before it is drained at the edges of the iris into the body's circulatory system, which happens roughly 12 times a day. Eighty percent of the volume in the back chamber is made up of a jelly-like blob, called the *corpus vitreum,* providing the eye with nutrients. This is essentially what gives the eye its shape and even contributes to the structural integrity of the skull. Illustrating this fact, we see how people who are born without one or both eyes often get cranial malformations.[18]

After its journey through the humours, the light finally makes contact with the retina. While the cornea and the lens focus light to the fovea to create a clear and crisp image, the rest of the retina is also bathed in photons — most of it in fact, since the central fovea is so small.[1] This is where the design also leaves some things to be desired. Let us do a thought experiment: say the light is a ball and the retina is a wall in front of you that you want to hit. In the middle of the wall, you are going to have a small circular region that is completely bare, roughly four balls wide. The rest of the wall, however, is covered in layers of crawling vine plants. You can see the wall behind it, but it is not going to be a straight shot. Also, one spot of the wall, roughly the same size as the bare central spot, is completely covered by a great old root, supporting the vine and making the wall completely invisible. If the ball you are throwing is covered in red paint and you were to throw it at the central, bare part of the wall, there would be a clear and distinct mark left by the ball. We can clearly see where the ball has hit, and judging by the round shape, we can probably tell it was a ball without knowing so beforehand. On the other hand, if you were to throw it anywhere else, there would only be a smidgeon of red visible through the vines, and while we could probably tell where the ball has been thrown, we would not be able to say what caused the red smudge. If you were to throw the ball at the root of the vine network, there would be absolutely no impact on the wall. This is essentially an analogy for how the eye works. The central part is that very sensitive fovea where light hits the photoreceptors directly, inhabited mostly by colour-sensitive cones. The vines covering the wall represent small nerve fibres covering the retina, called axons. These axons are what relay the information from the rods and cones, as the light is converted to electrical impulses. These axons crawl over the retina, leaving the eye together in *the optic nerve*, which is the root of these small axons, similar to the root of the vines. This is otherwise known as *the blind spot*. Just like the ball cannot hit the wall, any light falling on the optic nerve will not be picked up by any photoreceptors, so this region is completely insensitive to visual input. Thankfully, this blind spot covers two different regions in the visual field depending on which eye you are looking with, so the brain is able to fill in the blanks of one eye using information from the other.

So why, after millennia of evolution, do we only have such a small field at the back of our eye where we actually experience clear vision? Do the axons need to wander in front of the photoreceptors, blocking the light from entering, and then punch a hole through the retina when relaying the information to the brain through the optic nerve? While this is the best that human evolution has come up with, we will see later in the chapter that other animals have solved this problem much more elegantly.

Staying with human vision and higher visual processing, our axons pick up the electrical current created by the rods and cones as the light hits the disks. The network of roots meet up in the optic nerve, one of 12 cranial nerves, and leave the retina through the blind spot[1] (Figure 1). In truth, the optic nerve is not as much a nerve as it is an elongation of the brain itself, a feature it shares with the olfactory nerve which relays information of smell. For that reason, when your eye doctor is looking into your eyes, they can technically see a part of your brain. Therefore, there is certainly some truth to the saying "the *eyes are a window to the soul*", or rather the brain.

Creating an Image

Once light has been translated into electrical impulses and enters the brain, this is where the pathway becomes somewhat more convoluted. Seemingly unhappy with their allotted place in the skull, the cranial nerves switch over to the opposite side. This happens shortly after the nerves have entered the brain, in a crossover station called the optic chiasm (see Figure 1).

After the "Great Crossing", the signal continues through the optic tract, which is essentially a continuation of the optic nerves. Importantly, however, there has been a division of the electrified light. During the crossover in the optic chiasm, only 60% of the nerves actually entered the other hemisphere of the brain.[19] This means that 40% of the information actually stays on the same side as the eye from which it originated. There is a rather simple, almost mechanical reason behind this. While the optic tract received information from both eyes, it has now been divided according to where in the visual field the light hit.[1] To illustrate this, consider watching a bird nesting in a tree and you are facing it straight on so the light reflected from the bird will land on different parts of your two eyes. For your left eye, the light appears to come from the right, while the

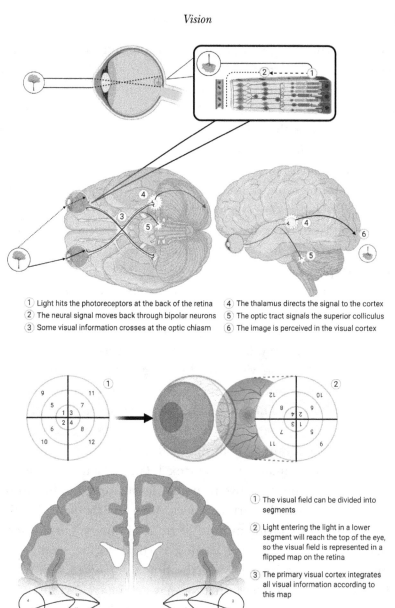

① Light hits the photoreceptors at the back of the retina
② The neural signal moves back through bipolar neurons
③ Some visual information crosses at the optic chiasm
④ The thalamus directs the signal to the cortex
⑤ The optic tract signals the superior colliculus
⑥ The image is perceived in the visual cortex

① The visual field can be divided into segments

② Light entering the light in a lower segment will reach the top of the eye, so the visual field is represented in a flipped map on the retina

③ The primary visual cortex integrates all visual information according to this map

Figure 1.   An image is formed. In the first image you can see how incoming light, in this case reflected of a tree, reaches the visual cortex of the brain. As illustrated, the image reaches the brain flipped due to the way light is conveyed through the eye. You can also see how the incoming light will hit different sides of the optic nerve in the two eyes due to its angle of trajectory. In the second image you find the retinotopic map, represented both in the eye and the brain. Created using Biorender.

opposite is true for your right eye. Still, the light comes from the same source, and to compensate for this, the signals somehow have to cross over, which the body has solved beautifully by simply making the nerves cross. For comparison, this is less elegantly handled in the albino tiger as they are forced to go cross eyed. This might be difficult to visualise, so the previous illustration in Figure 1 may help. In particular, consider how the trajectory of the light, though coming from the same point, will hit either the outer or inner part of the retina depending on which eye is closer. This means that information from the same part of the visual field can be processed in the same area of the brain, the visual cortex in the occipital lobe, which is responsible for visual processing. Ironically, this structure happens to be located furthest from the eyes. You could argue that the visual cortex is what we actually see the world with, and the eyes are just sophisticated tools through which the brain can reach the outside world. Arguably, this is true for all sensory systems.

Objects in the same visual field ending up in the same place in the visual cortex is not only true for left-and-right visual fields, but also top-and-bottom. This means our vision is essentially divided into four quadrants, or quarters. Thanks to the crossing of the optic nerves, these quadrants are all handled in the same region of the brain, known as primary visual cortex 1 (V1).[1] Generally speaking one might claim there are five principal visual areas, but for simplicity we will focus on this primary area. V2-5 mostly deal with finer analysis of what is moving around in front of our eyes and where it is heading, but it is V1, the earliest cortical region dealing with vision, that has the highest specialisation for pattern registration, and consequently produces what we may call image-forming vision. Still, this is a broad simplification, as vision employs a huge cortical and subcortical network in processing all aspects of incoming light.

The following description of how light is translated into vision is based on the findings of Hubel & Wiesel, which saw them receive the Nobel Prize in 1981. It deals with the fundamentals of image-forming vision, but vision enthusiasts may be disappointed by a lack of supplementary models. The fact is that vision, and all senses, are such a complex amalgamation of basic feature detection and perception; the latter deals with the perspective of cognitive neuroscience, aspects of which this book may sometimes refer to but for the sake of brevity sadly must leave to other works to describe in more detail. This caveat holds true for all chapters in this book, and I have included references for those who wish to delve deeper into the merging of sensation and perception. There are several books out there dedicated to individual senses. If we were to aim to adopt a more holistic approach in this humble introduction to sensory neuroscience, it would most likely force even greater oversimplifications to the extent that it may do injustice to the field. For that reason, I ask you to bear in mind this is an abridged version of the brain's inner workings.

Returning to our visual pathway, the primary visual cortex acts like a very advanced version of that first primordial eye in our first example, only being able to sense if something was light or dark. The brain similarly differentiates between light and dark, but with a much higher resolution that makes it possible to detect edges and outlines. This sensitivity to contrasts varies between the different neurons that inhabit the visual cortex. Such neurons have what is called *orientation preference*, meaning that different neurons react differently to the direction of the object being seen, whether that be vertical or horizontal.[1]

The visual cortex is also composed of columns. Neurons of a column will react differently depending on what is being seen. In a particular direction, these form *orientation columns* based on the neural signalling described above; in the other direction, we have *ocular dominance columns*. The latter depend on the strength of the visual signal from the left and right eye. If a stronger signal is received from the right eye, then that corresponding column will increase in size with formation of more synapses, or nerve connections; by contrast, the columns related to the other eye will decrease in size. While the orientation columns help us decide what an object is and where it is going, the function of the ocular dominance columns is up for debate. Many argue that it is important for bin-ocular vision, being able to fix on a target with both eyes, while others argue that it is meant to act as a relay station to shorten the signalling distance to other parts of the brain. Some argue it lacks a function altogether and is the brain's version of a seemingly redun-dant appendix.[20]

Essentially, the way we perceive the world is in terms of horizon-tal and vertical stripes, the strength, contrast, and direction of which are computed into forming an actual image; this is then stored in the visual cortex like a biological hard drive. So how much informa-tion are we talking about? If the brain is likened to a computer, then we should be able to translate this into streams of data. Let us take your internet service for comparison. An Ethernet cable is usually able to send between 10-100 million bits per second, with a *bit* being a single unit of information.[21] As the eyes relay about 10 million bits to the brain, we would consider it a low-tier internet provider. Thankfully, this does not bother us very much, considering we can only consciously be aware of about 50 bit per second.[21]

So what happens to that data we seem to be ignoring? Much of it goes towards controlling our bodily functions, realigning the eyes or controlling light-intake by adjusting the size of the pupil.[1] However, we can be made aware of some of this "subconscious" information. If you have ever driven a car, or ridden a bike, and completely zoned out going back from work or school as you are going down a familiar route, only to find yourself on your street with no idea of how you got there, you may appreciate why. This autopi-lot can take over because your frequently used daily commute has been imprinted in the neural network of your brain, and the visual information relayed through our eyes is just used to double-check this well-established framework meets what is expected.[22] With fre-quently repeated tasks, you are often not made conscious of your surroundings. Should something deviate from the brain-map, like the neighbour's dog running into the road, this new information would grab your attention, waking you up just in time to hit the brakes.

As well as forming a visual image, the neural signalling causes a series of subtle changes in the body.[1] For example, and as illustrated in Figure 1, before the visual signal hits the visual cortex it will have been relayed through the *lateral geniculate nuclei* in the *thalamus* which is responsible for relaying sensory signals as well as regulating consciousness and sleep. From here, a signal is also sent to the *pretectum* where the size of the pupil and shape of the lens is regulated; if the field of view is dim, the pupil will dilate, letting more light in. If the object viewed is near, the lens will become rounder due to the muscles controlling it slackening.

The things you see also affect the way you move, even without you thinking about it. The neural signal, relayed by the photoreceptors through the optic nerve, also reaches the *superior colliculus*. From here, a decision is made on how your head should move to accommodate what is happening in your visual field. This lets you reflexively move your arm to catch that glass of water tipping over, as well as coordinate your eye movements to keep up with the tennis game on TV. This all happens silently and subconsciously. Therefore, while the human eye has some flaws in design, its workings downstream are nevertheless impressive and reach out beyond simply vision.

To highlight this, our perception of night and day is also controlled by what we see. Just above the optic chiasm, where the optic nerves cross from one side to the other, rests the *nucleus suprachiasmaticus*. Here, the brain decides if it should consider it to be night or day, according to the circadian, recurring, cycle. If it is light, the brain will adjust itself as if it was the day-period of the cycle; in the absence of light, we perceive it to be night and time for sleep. This is all controlled through a series of hormonal releases into the body. As an example, you might have heard about how birds are tricked into believing it is night when a sheet of dark cloth is put over their cage. Essentially, our brains behave in the same way, though we are somewhat better at seeing through the ruse of a cloth being put over our eyes.

Vision and Perception

The complexity of the human eye becomes apparent by what happens when something goes wrong and they cannot work as normal. As you might imagine, having your visual system affected can cause some rather strange phenomena. We have already seen that photoreceptors in the retina help to establish both what things are, and where they are going, with the cones being good at identifying the outline of an object and the rods at identifying its direction. The brain does the same with the information it receives; from the visual cortex, there are two *streams* carrying precisely that information for further processing.[1] The *ventral stream* tells us *what* something is, while the *dorsal stream* relays *where* the visual motion is heading.[23]

The way in which we perceive visual information is the outcome of a complex neural network, with different parts of the brain contributing towards presenting a clear and well-defined image or event. All this complexity is of course highly susceptible to the way we interact with our surroundings and our preconceived notion of how the world is *supposed* to work, which affects how we infer meaning from visual input.

There are essentially two ways to deal with sensory information: *bottom-up* and *top-down*.[1] Bottom-up processing involves sensory information being received on the ground-floor, through our sensory receptors in the eyes, skin, and ears or wherever they happen to be, and then travelling up to the brain for integration and processing. In a way, this type of analysis allows sensory information to be relayed as it is naturally intended, unfiltered by any preconception. You see a baseball approaching you, and being the trained catcher you are, you catch the ball without even thinking about it.

The inverse, top-down processing, works differently in that the higher centres of the brain affect how you interpret a piece of information. This might manifest as an emotional interpretation of an image. For example, if you see two humanoid shapes which are identical in their rather angry posture, although one is slightly smaller, you might assume that the smaller one is located behind the bigger one due to their size differences, and the angry expression of the closest figure might be interpreted as frightened as it could be considered that it is being chased by its angry twin.

There are also anecdotes regarding how the way we grow up might affect our perception, but these should be taken with a healthy pinch of salt. For example, the anthropologist Colin Turnbull once described meeting a tribe of people living in a dense forest, rarely moving out into the open.[24] When the author accompanied a young man into the open fields, they came upon a herd of buffalo in the distance. The writer thought it best not to approach and to leave the giant beasts alone. The other man, however, saw no such issue. After all, these animals were hardly bigger than ants and could not possibly do any harm to them. What this story suggests, if it is indeed true, is that depth-perception is something we humans develop over time. Having grown up in the dense woods, the man had no experience of estimating the size of things on the horizon, with most visible things being within a few metres.

As you may notice throughout this book, we can often learn quite a bit about the function of a brain region by assessing symptoms from local injuries. A stroke in the small blood vessels supplying the neurons of the dorsal stream starves this area of oxygen and can cause a person to not be able to perceive visual motion. As a consequence, that person will view the world as a series of static images, like photographs. When pouring tea into a cup, that person would first see the teapot being tilted towards the cup and then suddenly be greeted with an image of having spilled tea all over the table, as the slow rising of the tea in the cup is not being relayed to the right centres of the brain. If, on the other hand, the stroke hits the ventral stream, the symptoms would be quite different. In this scenario, someone will see motion and can detect shapes and colours.

However, they will be incapable of putting them into a setting, identifying *what* it is they are seeing. This is called *apperceptive agnosia*, with agnosia referring to an incapacity to detect certain aspects of a surrounding.[25] In this example, a patient with apperceptive agnosia can pick out a teapot from a series of different pictures, but will not be able to see that it is a different shape from a potted plant, for example.

While the patients above cannot identify shapes or motion, damages to the temporal lobe which houses the visual cortex will lead to a completely different set of symptoms. This might come in the form of *associative agnosia*.[26] These patients can recognise an object and describe the shapes of different parts of it, but will not be able to identify what it is they are describing. A circular object with a semicircular handle and a protruding spout with a dark liquid running from it is just a series of shapes, and while they know what a teapot is, they will not be able to identify the object they are looking at as one. Even in front of a mirror, seeing a nose, ears and eyes, patients with associative agnosia might not be able to tell that the shapes they are seeing make up a face, even though it belongs to themselves. A variety of this disorder is called *prosopagnosia*[27] (face blindness) where a patient cannot make out faces specifically, as described in Oliver Sacks' popularised portrayal in *The Man Who Mistook his Wife for a Hat*.

Clearly, human vision allows for some interesting mistakes. As such, it is hard to claim it could be the perfect pinnacle of evolution some present it as. Still, despite its imperfections and flaws, it makes up a cornerstone of what makes us human, shaping our cultures and experiences. As we have touched upon in this brief introduction to visual neuroscience, the system can be tricked, and vision is not necessarily the same as perception. Everything we see is coloured by preconception; necessarily so, since we cannot afford to relearn the dangers of a busy intersection every day. The downside is that it also spills over into other aspects of life, and shifting one's own perspective is often quite difficult. While we may aim to rationalise our opinions and behaviours from the perspective of others, it can be helpful to remember that even though the same events may cross our retinas, our perception is dependent upon a neural structure that can be influenced by individual differences. The Canadian psychologist Donald Hebb once said that "neurons that fire together wire together", echoing the impact of upbringing and environment on our very physiology. With that being said, vision connects us far more than it separates. In fact, the way nature has created a variety of eyes through different evolutionary pathways is truly inspirational, as we shall see in the next section on eyes in the animal kingdom. Separated by eons, across species, we have found different ways of interpreting the same photons, seeing the same things as our neighbours, though sometimes attributing them with different meanings.

Having gone through the human solution to the visual puzzle, we are somewhat better equipped to evaluate our sensory proficiency. Does the eye then, with all its complex biological machinery, live up to its reputation as the epitome of evolution, or is it instead a case of "not all that glistens is gold"? Certainly, there is beauty in how they allow the brain to create images, but one could also argue that the eyes have been subject to plenty of human-centric propaganda. Sure, our ocular protrusions are amazing, capable of relaying visual information of the world around us. Still, had the eyes been a school project, they would probably have been sent back for revision. I dare even suggest there are some fundamental flaws that could do with some improvements before we start modelling cybernetic implants based on the organic visual organs we have today. Nevertheless, vision constitutes one of the most important tools of biological life. Human vision might not be perfect, but it has certainly gotten the job done. It does however leave much to be desired, as we will certainly discover when we compare ourselves to some other animals out there.

Chapter 3

Hearing

Good Vibrations

I remember the launch of a new model of earphones by a very prominent tech company. The CEO, speaking on stage in front of thousands of fans, made the compelling argument that their new earphones were designed to direct sound into your ears, as opposed to their competitors' earphones presumably blasting music into the surroundings. What the CEO meant was of course that the design of the earphones was aimed at focusing the sound in a more direct way, reflecting the complex anatomical structures that allow your brain to enjoy music, by simply interpreting the frequency of vibrating air.

That is what hearing all comes down to in the end: vibrations. Due to our ancestors' decision to emerge from the oceans onto land, these vibrations travel by air for all land species. You may have noticed, however, that sound carried through any other medium, like water, can be heard even by us. The reason for this is not that air has some magical capacity for sound. After all, dolphins and whales make perfect use of their watery surroundings, squeaking tales of the depths to any who might care to listen. The reason air is needed to produce intelligible sounds for us humans is that our anatomy has developed specifically for it. Essentially, it is a simple matter of converting energy according to Newton's third law of motion. For example, if you hit two metal bars against each other, you will feel the vibrations in your hands due to the conservation of energy. While the bars will carry most of the energy away, all matter that is in contact with the collision will be affected. Even though we cannot see air, the molecules it contains cannot escape Newtonian laws and will absorb some of the energy from the blow. This energy is what causes the vibrations we refer to as soundwaves.[28]

First, let us explore those vibrations or soundwaves. Sound is nothing but pressure waves composed of vibrating molecules. So wherever molecules exist, there can be sound, which pretty much only rules out the vacuum of space. Just like waves on water, the strength of soundwaves dissipates over time. This is, for example, why the wonderful but rather loud art of yodelling was invented, aiming to communicate messages over the large distances in the Alps.

The height, or amplitude, of the wave, is what decides how *loud* the sound will be and is measured in *decibel (Db)*. If you happen to play a musical instrument, you may also be familiar with the concept of sound having a certain frequency. The frequency, measured in *hertz (Hz)*, tells us how many waves pass by during a second. The more waves that pass by, the greater the *pitch* of the sound. Generally speaking, humans can hear sounds between frequencies of 20–20,000 Hz.[29] Younger children can often hear pitches that go even higher, whereas adults can only perceive an upper limit of 15–17,000 Hz. I remember this concept being illustrated during a class in high School: someone had recorded a high frequency sound on their phone, setting it off quite loudly during the middle of history class. While all the students looked around for the source of this high pitch screeching, the teacher just looked on perplexed. We had quite a hard time convincing him that there indeed was a sound. Despite a sound being loud, it is its frequency that determines whether our ears will detect it or not.

An Inconspicuous Mutation

So, knowing now what sound is, let us rewind the clock a few millennia. While sound has existed for as long as Earth has had an atmosphere, there certainly was a time when no living creature could hear any of those primordial echoes. To any inhabitant, Earth would have appeared completely silent. So why and how was it that some primordial creature thought it wise to upset that unending serenity? Compared to many other aspects of our physiology, hearing seems to have joined the stage relatively late, when fish already had well developed gills. One of these fish, the Eusthenopteron, seems to have developed an initially rather inconspicuous mutation.[30] A malformation of a piece of the lower jaw had gotten in the way of its gills. This bone, is called a *spiracle*, derived from the Latin word *spiraculum*, meaning breathing. The spiracle seems to have been a key factor in allowing its owner to breathe air while swimming on the surface. On top of this, this bone also allowed fish to sense vibrations in the surrounding water. To this day, the spiracle continues to play an integral part in human hearing, as it evolved into the stapes or stirrup bone, one of the small bones of the human inner ear. Back in the sea, fish now possessed what we may very well consider an ear. While not exactly capable of enjoying the latest musical hits, this ear allowed the identification of movements in the water through perception of vibrations. A fast and loud vibration might have signified that a smaller fish was passing by, flapping its tail wildly. Similarly, a slower and weaker vibration told a story of a large predator swimming in the distance. The speed and the strength of the vibration could therefore tell primordial listeners about what had made a particular sound, helping it decide whether to approach a potential meal or escape to avoid becoming one itself. While this interpretation of waves might not feel like true hearing in the human sense, it most certainly matches its evolutionary purpose in allowing the listener to avoid danger. Today we might be more likely to use our hearing to jump out of the way for a bus rather than swimming away from a shark, but ultimately, it all boils down to the same general principle: our instinctive reaction to intense and approaching vibrations that we have learned to associate with danger.

Moving from water to land is no small feat but clearly the primitive hearing apparatus was successful enough to be conserved throughout evolution, with bone-led vibrations being used by both species hearing through air or water. We will go into greater detail on how differences may manifest in the anatomy of terrestrial and aquatic animals, but let us first establish the basics of how these bones work. It all centres on *impedance matching*, which essentially states that in order to maximise the transferral of energy, the resistance in the incoming signal needs to match that of the receiving end. In electronics, this can be ensured through adding resistors within circuits. When it comes to hearing, bones with specific properties that allow them to move with just the right amount of freedom act as the equivalent of resistors. These bones essentially translate the incoming vibrational energy into a signal the brain can safely and reasonably integrate. Without a hearing apparatus, vibrations might reach the brain anyway, especially in fish where the vibrations go from liquid to liquid; however, the lack of a dedicated energy relay would make these vibrations difficult to make sense of. This need for impedance matching is even more critical in humans, where vibrations go from a medium of low (air) to high (liquid) conductance. The evolution of hearing from the primitive apparatus of the malformed gills in early fish to the much more complex structures of the human ear was a long evolutionary journey. Although this generated many different types of ears, the core principle of impedance matching remained consistent.

The Human Ear

Early humans, while certainly possessing more complex ears than the Eusthenopteron fish, unsurprisingly looked quite different from modern man. Based on archaeological findings, computer models have recreated what an early human ear may have looked like. Most notably, this has been done with the remains of the Australopithecus africanus and the Paranthropus robustus, both of whom made Africa their home about 2 million years ago. These primates would have had an easier time hearing hard consonants like T, K, F, and S, in the range of 1,000–3,000 Hz than modern humans.[31] It is believed these sounds would have carried well across short ranges over open areas, such as the open plains of Africa.

Humans eventually mastered the technique of manipulating airborne vibrations by rapidly closing or opening their trachea, or windpipe, with the air escaping used to produce different sounds and eventually speech. Most evidence points towards the first forms of verbal communication having developed in Africa some 100,000 years ago.[32] These languages would have emerged with respect to the sounds early humans could easily hear and decipher. This would most likely have been in the form of click-languages,[33] producing distinct and clear sounds much like the T, K, F, and S sounds described above; it is still debated whether these constitute the truly first human languages. Even today, this ancient way of speaking has been preserved in languages still widely spoken in Africa and I would encourage everyone to give learning some Xhosa or Zulu a go.

Now, after having gone on about the wonders of the human physiology that allows us to hear, let us explore how human hearing actually works, an illustration of which you can find in Figure 2. Unsurprisingly, we will start with that peripheral system which allows the body to direct airborne vibrations into the otherwise well protected skull: the human ear. Interestingly, the outer ear or the part visible from the outside makes up only one third of the actual ear.[34] This region is then followed by the *middle ear* and finally and unsurprisingly, the *inner ear*.

The outer ear itself is comprised of three parts. The part of if that you see, the *auricula,* makes up the wing-shaped cartilage structure framing your head. The more rigid part of the ear that surrounds the entry into the skull is called the *concha,* which in turns give way to the ear canal, or the *meatus aucusticus.* These three parts share one important function: to convey soundwaves and focus them onto a small membrane in the middle ear called the *membrana tympanica, or more colloquially known as the eardrum.*

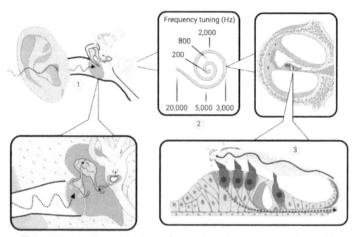

1. Sound hits the eardrum and the vibrations are carried through the three miniscule bones of the inner ear
2. The frequency of the vibrations dictates where in the cochlea they will cause the strongest response
3. The basilar membrane is attached to small hairs that move with the vibrations, creating a neural signal

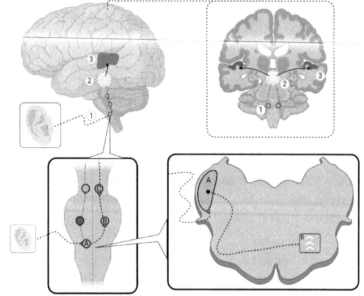

1. The neural signal enters the cochlear nucleus (A), mostly crosses over to the superior olivary nucleus (B) and later the inferior colliculus (C)
2. The thalamus directs the sound signal to the relevant brain regions
3. The auditory cortex allows us to perceive and conciously interpret the sound

Figure 2. Making sound. The first image depicts how vibrations in the air is conveyed through the inner ear in order to create an electrical impulse. The second illustration shows how that signal travels through the brain, allowing us to interpret the sound as something meaningful. Created using Biorender.

However, before sound reaches the eardrum, the vibrations are manipulated by the outer ear. On their initial contact with the wing-shaped auricula, the soundwaves will enter the ear canal differently depending on their direction. The folds in the ear make it so that a sound reaching the ear from above is amplified compared to sounds coming from below. As a result, we can easily tell the vertical direction of a sound, with a clear proclivity towards prioritising threats from above.[35] One can certainly theorise on why this might be the case. It is unlikely that our ancestors, albeit smaller, were overly concerned with large predatory birds aiming to whisk them away. The greatest threats are, after all, on the ground: lions, bears, hyenas, and most importantly, snakes. Perhaps this form of directional hearing functioned instead as a social relay-network. For instance, let us consider a gathering of apes, resting while safely tucked away in a group of trees just on the outskirts of the savannah. Resting on some low branches, adults are watching as their young ones play on the ground. Suddenly a call rings out. A young adult, having tested his strengths by climbing up high into the crown of the tallest tree, has spotted something: a pride of lions is approaching and they look hungry. Before the children can understand what is happening, they have been forcefully snatched up by their mothers and carried away; they will have to wait some million years before the cats are small enough to be safely played with. It is reasonable to postulate therefore that our primate ears help us navigate through sensory blind spots. For instance, we can see what is happening in front of us using our visual system, but a panther hiding in the trees above us would go forever unnoticed if we were to rely solely on our eyes.

Returning back to our ear, having been directed by the folds, the soundwaves then hit the ear canal. Due to its length, this narrow path has its own frequency of 3,000 Hz, meaning that any vibration entering this narrow tunnel will be modified to fit that mould.[19] As a consequence, soundwaves close to this frequency, in the range of 2,000–6,000 Hz, will be amplified 30 to 100 times. This gives humans an apt sensitivity to sounds in this range, a range which corresponds to human speech. Naturally, this is far from a coincidence. Human speech is not set at around 3,000 Hz because of the length of the eardrum. Rather, the eardrum and human capacity for speech developed in such a way so as to complement the work of its owner's vocal cords.

Taming the Airwaves

As we reach the eardrum, the first part of the middle ear, the sound-waves are up for somewhat of a surprise. As the airborne vibrations are prepared for being translated into nerve impulses, they will have to pass by a wall of water. As the soundwaves hit the eardrum, they are quite literally translated into water-waves in the liquid behind the drum. As water has a much higher resistance than air, most of the energy in the soundwave is reflected, and consequently lost, on contact with the eardrum. In fact, 99.9% of the energy is lost on this first contact, filtering out most of the incoming vibrations.[19]

The energy that does carry over is however immensely magnified through impedance matching,[1] allowing the pressure imposed on the eardrum to increase 200 fold before reaching the inner ear. There are two essential components to this principle. Firstly, the entry into the inner ear, the *fenestra ovalis* or oval window, is 20 times smaller than the eardrum. Consequently, the energy is directed to a much smaller area, thereby increasing the force with which the waves travel. Furthermore, the three small bones in the middle ear not only help to convey energy but they also amplify it.

The first of these, the *malleus* or hammer, transfers the vibrations from the eardrum to the *incus*, the anvil.[1] This is where that little malformed jawbone of our ancestor fish comes into play. That initial spiracle, sensing vibrations over the gills, is retained in modern humans in the shape of the third middle ear bone, the *stapes* or the stirrup bone.[30] This is the smallest bone in the human body and it connects the anvil to the oval window.[34] These three miniscule bones act as a series of levers that allow the initial vibration hitting the eardrum to be magnified on its route to the oval window and the inner ear.[1] Thankfully, there is a safety mechanism in case the incoming sound is too great, at least to some extent. The *tensor tympani* and *stapedius* muscles are attached to the eardrum and stirrup bone, respectively. If there are intense vibrations, these muscles contract, reducing the transfer of energy and therefore the volume of the sound perceived.

The oval window connects to the final part of the ear. This inner ear contains one major structure that is responsible for translating the incoming vibrations into distinct nerve impulses that the brain can then comprehend: the *cochlea*.[34] This bony labyrinth is curled up, much like the shell of a snail. In this configuration, it is about 10 mm wide but rolled out it would be over 3 times bigger. Three hollow structures fill the cochlea and they are filled with different types of fluid that play an important role in transferring the mechanical energy of the vibrations into electrical nerve impulses.

The hollow chamber that is attached to the oval window and which is in direct contact with the middle ear is the *scala vestibuli* or the vestibular duct.[1] As the membrane in the oval window starts to vibrate, physical waves form in the vestibular duct in a type of fluid called *perilymph*. These waves will continue through the labyrinth towards the tip of the shell, as seen in Figure 2. Here, the waves continue in the second duct, the *scala tympani*, or the tympanic duct. This duct continues all the way back to the base of the cochlea and ends just below the vestibular duct and the oval window, in the equally well-named round window. In principle, you can imagine this structure as a tube with membranes covering the opening on either side. The membrane on the far end, the round window, allows the energy carried by the wave in the cochlea to escape rather than spreading chaos in the tiny space the perilymph resides in. So when the oval window moves inwards, the round window will move outwards, creating a distinct motion inside the bony labyrinth. It is this motion that ends up translating sound waves into nerve impulses.

The third, and last, chamber in the cochlea is the *scala media*, or cochlear duct, nesting between the vestibular and tympanic ducts, but never connecting to them.[34] While the cochlea itself is composed of bone, the inside is divided by thin membranes, creating the three ducts. As a result, the inside of the inner ear is not very stiff, instead allowing for movements which are essential to the production of sounds.

As the wave travelling through the perilymph of the vestibular duct comes in contact with the basilar membrane, the energy is once again transferred in the form of vibrations.[1] Depending on the frequency of these waves, different areas of the membrane will begin to move. The basal part of the membrane is thinner and stiffer than the regions closer to the cochlear tip which are broader and more elastic. Considering that the wave will always reach the thin region first, this is where the vibrations are carried over and they spread through the membrane as a *travelling wave*. As the membrane becomes more elastic, the size or amplitude of the wave will increase while the frequency decreases. Therefore, each type of vibration will result in specific maximal amplitudes with set frequencies. This peak amplitude will occur at different places in the basilar membrane depending on the frequency of the original soundwave. Simply put, a high frequency sound will produce a peak signal in the basal part of the membrane, while a low frequency sound ends up in the more elastic apical region. This creates a sort of sound-map inside the cochlea.[1] You can think of the basilar membrane as a sort of piano: depending on where you push down the keys, a different note will be produced. This sound-map, or tonotopic map, continues up the chain of command into the brain.

We have talked quite a bit about vibrations, and at this point, one might expect the trend to continue, with sound just being the by-product of our brains' vibrations against the inside of our skulls. Thankfully, that is not the case. The basilar membrane breaks the chain of vibrations, translating the mechanical energy first into chemical, then electrical signals. This is done with the help of tiny hairs called *stereocilia*. While small, they are quite numerous as 15,000 of them compete for a spot in the narrow ducts in the cochlea.[36] The bodies of the hairs are firmly attached to the basilar membrane and reach into the cochlear duct where the tips are anchored to a second structure, the tectorial membrane.[1] The vibrations of the basilar membrane cause the cilia to move. This, however, is where that perpetual vibration dies out. The tectorial membrane is rigid. As the cilia rock back and forth, the mechanical energy is finally converted to electrical impulses. This transduction is ultimately triggered by *tip links* connecting the tips of adjacent hair cells. When these links are stretched by the movements of the cilia, small channels are opened in the hair cells. As positively charged atoms, or ions, enter the cilia, the hair cell changes its electric charge. As a result, the little vibrating hair cell now carries an electrical charge itself. As the cilia move back and forth, matching the soundwaves hitting the basilar membrane, the channels within their walls open and close, depending on the direction and the tension in the tip links. The flow of charged ions in and out of the hair cells causes an electrical charge which will be of different strengths depending on the position of the hair cell.

Much like our ears, the cilia have developed certain traits to fulfil specific functions.[1] There are two essential types of hair cells: *inner* hair cells that respond to vibrations of the basilar membrane, and also a group of *outer* hair cells. The outer hair cells are almost completely controlled by the brain, and like undercover policemen, they aim to blend in with the inner hair cells. They do this quite well and vibrate with the same frequency as the soundwaves, allowing the brain to detect the frequency of the incoming signal. The movement of the outer hair cells in tandem with that of the inner group strengthens the active movement within the cochlear duct, much like a built-in amplifier.

Let us focus a bit more on the electrically charged inner cilia. Before the brain can register the frequency of their signal, they have to relay this information somehow. Each hair cell is attached to a nerve and as soon as the charge is built up in the cilia, an electrical impulse travels through the hair and on to this nerve[1]. Due to the "1:1 hair cell-nerve" ratio, this signalling pathway is much more sensitive than the much blunter system of the eye where several rods may have to commute through the same neural highway. The downside however is that each nerve fibre, or axon, can only relay a very limited part of the audible spectrum.

As mentioned, different parts of the basilar membrane vibrate depending on the original frequency of the soundwave. Naturally, this affects the hair cells much in the same way and the nerve fibres in the beginning of the cochlea will fire their electric charge in response to high frequency sounds. The fibres connecting to hair cells in the tip of the cochlea will instead produce a signal in response to lower frequencies. The nerve signal is however not always 100% in sync with the dancing hair cells. For frequencies below 3,000 Hz, it is a rather straightforward process, with the electrical signal and the frequency being the same. For higher frequencies however, the ear needs some additional support.[37] As a result, the tonotopic map produced in the brain needs to be consulted to aid interpretation of higher frequencies. By interpreting the different charges, the brain can draw more precise conclusions than the inner ear alone, all without you even noticing it happen.

The hair cells do however provide some additional help in interpreting the signal, even without the brain's interference. The hair cells that dance to the higher frequency are pickier about what tunes make them move compared to the cilia in the cochlear tip. This narrower window of sensitivity means that rather small changes in frequency will cause a certain group of cells to stop and others to trigger.[1] As a result, we find it easier to detect differences in pitch for higher notes. Furthermore, when the cilia have been dancing for a very long time due to a constant note being played, they will stop producing an electrical signal. The brain simply stops getting a signal of that specific note, in a process of adaptation. Much like you can filter unwanted sounds from a digital recording, the brain filters away background noise this way all the time, a property that some medical treatments that we will discuss later use.

Vibrations Become Sound

Leaving the schematics of the inner ear behind, let us move on to higher echelons of the auditory pathway, also laid out in Figure 2. After having been produced in the tiny hair cells and passed on to the nerve fibres, the electrical signals enter the vestibulocochlear nerve.[1] This eighth cranial nerve is a joint highway-to-the-brain for both hearing and balance, the latter of which is relayed through the vestibular system but we will deal with that in another chapter. Luckily, the nerve is sufficiently organised so that the sound signal can be safely isolated and ends up in the cochlear nuclei in the brainstem. These are the ventral and the dorsal nuclei, which anatomically point towards your throat and your neck, respectively. The ventral nucleus can generally be considered as the big brother of its dorsal counterpart. The tonotopic map, allowing you to distinguish the pitch that was established in the cochlea, is repeated in this ventral area, from where the map will be passed on through an intermediary in the superior olivary complex before getting back to the cochlea. This is the feedback loop we discussed earlier, where the outer hair cells vibrate with the same frequency as the inner hair cells that initially produced the sound.

The dorsal cochlear nucleus, by contrast, receives its sound information second-hand from the ventral core. While the ventral nucleus generally deals with the actual content of a soundwave, the dorsal nucleus is more interested in where the sound came from by integrating signals from other parts of the brain, therefore creating an internal map of its own position in relation to incoming vibrations.[1] The value of this integrated cartographer-of-the-brain should not be underestimated. You may be familiar with how some animals, such as bats and dolphins for instance, make use of sonar to create a mental image of their surroundings. In a very rudimentary way, humans also use our hearing to orient ourselves in relation to our surroundings. For instance, after some quick internal math, you can safely say that the fire truck is much further away than the mosquito buzzing by your ear, even though the latter is louder. There is a built-in system within our hearing network that gives some credit to this somewhat surprising comparison to sonar. As the firing of nerve cells is dependent on the direction of the wave travelling through the cilia, it is possible to compare the firing pattern between the two ears.[1] If one ear is out of sync with the other, then there must be a clear difference in distance between the two ears relative to the source of the sound. The time between the activation of the two ears helps us decide which side is closer. Similar to how we use both our eyes to achieve visual depth perception, we use our hearing to sense distances through sound. The decibel level of the sound only really comes in for frequencies over 3,000 Hz, due to the tonotopic map in the brain needing to be activated to tell the pitches apart. If it is an unknown sound, we will probably conclude that the ear receiving the stronger signal is the closest to the source.

What makes the distinct mapping of sounds possible is the organisation of the olivary complex. The medial part of the superior olive contains neurons that stretch in two separate directions, essentially towards either ear. Quite logically, the nerve that stretches towards the left will receive input from the left ear and the inverse is true for the right one. This holds true for both sides of the brainstem, as each olive complex contains two superior olive nuclei, one related to each of the two brain hemispheres. The superior olive nuclei in turn contain different segments. The neurons in these different segments act somewhat like random number generators which are activated when the two electrical signals reach them simultaneously. This is of great importance in allowing us to determine where a sound is coming from and can actually be explained by the way the nerves are wired. Quite simply put, in the medial segment, the nerve transferring the signal from the left ear is much shorter than the one coming from the right. This length slows down the electric impulse by exactly the same amount of time that it takes the soundwave to reach the other ear. As a result, the signal will be received simultaneously regardless of which ear the soundwave reaches first. If the sound had come from the right, a different segment of the superior olive nucleus would have been activated in the same way. That means that while there will be neurons firing left, right and centre, only one segment will be reached by two signals at the same time. When that happens, that segment acts like an ON-switch, letting us know the general direction of the sound. As mentioned earlier, this is only true for sounds below 3,000 Hz. For vibrations with a higher pitch, it is instead the intensity, or strength, of the sound that matters. The lateral segments of the superior olive nuclei take it upon themselves to deal with this.[1] In this respect, the actual neural wiring is ingeniously simple: the left side of the brain will receive an ON-signal from the left ear through the cochlear nucleus and an OFF-signal from the right side. These signals correlate with the intensity of the sound, so if the ON signal is stronger than the OFF signal, that simply means that the ear closer to that superior olive segment is the first to detect the incoming vibrations. Since this happens for both hemispheres, the matter of locating the sound is as simple as seeing which side is getting the stronger ON-signal.

Much like the tonotopic map helps us deduce the frequency of a sound, there is an internal map of localising the direction of a soundwave.[1] Once again, this happens higher up in the neural hierarchy as we bring the sound signal further into the brain. As seen in Figure 2, the nerves cross over at the brainstem after the olive nuclei and reach the ribbon-shaped lateral lemniscus. This structure is great at sensing the temporal aspects of the incoming sounds, meaning that it detects when a sound starts and when it stops. This is also where the acoustic startle reflex is initiated, where sound over 80 db will cause the lateral lemniscus to activate motor neurons controlling your limbs.[38] So when you jump because of a car horn urging you to cross the street faster, it is your lateral lemniscus in action. Interestingly enough, if a person were to have a stroke in the medial lemniscus, they would no longer be able to detect vibrations. Considering that sound is nothing but airborne vibrations, it is tempting to imagine these two segments at one point having been one, but ultimately deciding to sub-specialise.

Finally, the sound signal makes its way to the *inferior colliculus*. While the superior colliculus dealt with vision, its inferior counterpart ultimately creates the map allowing us to localise incoming sound.[1] This leaves one major pathway left for the soundwaves to take, which arguably is the most important one. Much like how one of the colliculi deals with vision and the other with hearing, the geniculate bodies have a similar relationship. In this case, it is the medial geniculate body that relays sound information. The sound signal leaves the inferior colliculus and travels into the medial geniculate body which makes up the auditory part of the thalamus. The thalamus essentially acts like the telephone operator centre of the brain, relaying crucial information to the places they need to be.

In the thalamus, the signals from both brain hemispheres join up. Their location, frequency and duration are compared between both pathways and made into one understandable signal.[1] From the thalamus, this signal finally reaches the auditory cortex. The first, or primary, part of the cortex also has the same type of tonotopic map we have encountered twice before while the signal was making its way up. If you think of a brain, you may visualise it as different parts lighting up as they are activated depending on the frequency of the sounds. The complex way these regions are activated is what makes it possible to hear so many different things at once but still makes you able to focus on a single voice in a room full of people. The specific pitch and intensity is put into context, and even with our eyes closed we would very likely be able to detect the voice of a friend, as well as his or her mood, simply by letting our brain put the pieces of the puzzle together.

I Hear What You're Saying

The original soundwaves, travelling as vibrations through the air, have made an amazing journey through different parts of your body, transferring from air to liquid and to electrical impulses before being translated into something completely different. The frequencies of these initial vibrations are what differentiates Mozart from a car horn and an ambulance siren from the sound of the ocean, and what makes you instinctively recognise your friend over the telephone. In a way, hearing may be thought of as a little bit like a highly specialised form of touch: the vibrations created by a raindrop falling on a tin roof physically impact the air around it, which eventually makes the air move against the tiny drum of your ear. So when someone says they have been touched by music, you could argue that to be literally true.

Air is however not the only means through which music may travel. In fact, it does not even have to pass through the outer ear, instead finding a different route to its inner sanctum. Most likely you have heard of Ludwig van Beethoven, the brilliant German composer whose work still inspires people almost 200 years after his death. You may also have heard that he went completely deaf, with his hearing starting to deteriorate as early as his early twenties. So how does one create such majestic symphonies without being able to hear them? You cannot, is the answer. Sound is, as we have established, just variations of vibrations reaching your brain. With his middle ears failing him, Beethoven found an alternative route for perceiving soundwaves.[39] Instead of relying on air, he would attach a stick to the keyboard of his piano and bite down on it. The vibrations would then carry over from the piano into his skull. These vibrations, albeit not as efficient as those carried via the air, would reach the inner ear, allowing faint sounds to be heard. Clearly, this was enough to create some of the most enjoyed musical pieces ever created. This same concept is interestingly even used today, for example when fitting patients suffering from middle ear deafness with hearing aids that convey soundwaves through the skull bones rather than air.

While the neural pathways that allow us to hear are sophisticated in their design, our human hearing can nevertheless be considered almost annoyingly delicate. For example, most people are aware that prolonged exposure to loud sounds can cause hearing deficits. This highlights the changing nature of our environment to the one we initially evolved in. For instance, the loud sounds we expose ourselves to today were not faced by our ancestors, and as a result our auditory pathways are not best equipped to deal with prolonged exposure to them. The problem with loud sounds is the impact they have on the very sensitive transduction of vibrations in the inner ear. To give this some perspective, one electrical discharge is generated every 0.000001 seconds as the small hair cells move 0.3 nanometres[19], or the diameter of a gold atom. That sort of speed is pretty impressive in a biological system and offers us an amazing range of hearing. As a point of reference, the Japanese earthquake of 2011 shortened Earth's day by 1.8 microseconds and that is almost twice the length of the hair movements. Just like shaking a tree might make leaves fall off, strong vibrations of the inner ear hairs will cause them to lose contact with the body. Since the hair cells lack the ability to regenerate, this causes irreversible hearing loss.

Hearing loss seldom comes alone after being exposed to dangerously loud sounds. Tinnitus is also an all too common reality for many people. Tinnitus is characterised as a constant ringing in the ears, often described as the kind of sound an old TV screen or radio might emit. The cause for this disorder lies in the outer hair cells of the inner ear. As we saw earlier, the inner hair cells send a signal to the brain, which returns that signal back to the inner ear, causing the outer hair cells to dance along to the tune of the inner ones.[40] This increases the impact of the original vibration and generates a sound which in itself can be identified with very sensitive microphones. These sounds are called oto-acoustic emissions and they cause the ringing sensation in tinnitus, due to loud sounds causing the outer hair cells to perpetually vibrate. Still, as with many other sensory disorders, it is difficult to pinpoint only one cause, and there is yet much to be learned about the condition. For example, it has been found that taking aspirin might increase the ringing, though it is unclear why.[41]

While we do not know of any way of salvaging these faulty hair cells, there are ways of limiting their unwanted effect. We previously said how the hair cells can adapt, filtering out background noise. This is also true for the brain. Much like you might get used to a scent, the brain will eventually ignore a tone that does not change its character, unless it is an oto-acoustic emission as in tinnitus. However, if we expose the brain to a tone of the same frequency as the one heard in tinnitus, the brain will learn to ignore that particular frequency. While this does not cure the tinnitus, it might go a long way in alleviating the worst of the symptoms[42] and is the reason why white noise is often trialled to help treat the disorder.

We have seen how our internal tonotopic maps are integral in locating the origins of a sound. In its most sophisticated forms, this phenomenon allows for one of the most valuable uses for hearing: echolocation. Characterised by an animal making sounds and judging the space around them based on the reflected soundwaves, this sensory art form is practiced by a wide range of animals. Before we start praising the internal mapping systems of echolocating animals such as the dolphin and the bat, we should take a moment to appreciate our own human capacity, as the brain clearly makes some effort to locate the origin of a sound. In comparison to vision however, this information becomes somewhat secondary when it comes to orienting ourselves. However, while this holds true for those with intact vision echolocation has become a viable method of adjusting to otherwise invisible obstacles. The principles of this are quite easy to demonstrate. You may do so by standing eyes closed in front of a wall. As you make a long shushing sound, the soundwaves will be deflected against the wall and you will hear the echo. Move one step closer, repeating the procedure, and the sound will now have increased. Move even closer and you will find that the sound has changed even more. With some training, you could probably become pretty adept at discerning the distances in your room using this method. Some people have taken this concept to impressive levels, and it would be unfair not to name some notable advocates of human echolocation by name as they have done much to inspire people all around the world. Daniel Kish, founder and president of World Access for the Blind, uses a technique he calls *FlashSonar*.[43] Combining clicks produced by his tongue and cane, he leads blind teenagers hiking and exploring outdoors.

Hearing then, much like vision, is simply a way of interpreting incoming forms of energy. Studies in people who use echolocation, like Daniel, have shown that their visual cortex will become when they interpret incoming sound.[44] Essentially, these amazing human echolocators train their brains to allow them to interpret the world around them without the aid of vision. Daniel even describes being able to make out ornaments on buildings as well as tell a wooden fence from a metal one based on their general outlay.[45] Ben Underwood, born in California, similarly became somewhat of an internet phenomenon as videos surfaced online of him skateboarding through his hometown, swerving around obstacles with the aid of echolocation.[46] Through making the seemingly impossible an everyday reality, these inspiring individuals demonstrate what the human body is capable of through determination and training.

Hearing also illustrates how our biology has changed since the onset of the first human languages. While clicks are still an integral phonetic aspect of several languages, there has been a clear evolution of both languages and people, intertwining culture and biology in a brilliant showcase of humankind's ongoing journey. Were we to bring back a human from 200,000 years ago, they would certainly not be able to understand English, but even the sounds themselves would be transported into the tunnels of their primordial ears and they would be able to decipher basic characteristics of the sound.

Chapter 4

Smell

An Early Sense

I once overheard a conversation between an overly confident author and his colleagues. The discussion was on the future use of robotics, and how they could or could not influence their realm of the arts. "Robots," he claimed, "will never be able to smell a rose". The sense of smell, more so than vision and hearing, seems to have often been perceived as being almost mystical. Most people know that we see light entering our eye, and hear things because of our eardrums moving, but one might find it harder to pinpoint exactly what constitutes a smell. Naturally however, smell is composed of something highly tangible and readily quantifiable by modern science and contrary to the beliefs of the previously mentioned author, there already are computers out there that are capable of identifying the scent of a rose.

The capacity to smell is called olfaction. Before we carry on with our odyssey through the sensory pantheon, I should clarify that we separate smelling into *orthonasal* and *retronasal* olfaction. The former deals with sniffing odour molecules and is more often than not what people refer to when talking about smell, while the latter is more involved in the more complex phenomenon we call *flavour*. For simplicity, this chapter will deal with orthonasal olfaction and we will return to flavour later when covering our sense of taste, as I believe most readers will find that grouping more familiar.

Let us start with our non-robotic selves and talk a bit about how the protruding piece of cartilage occupying the better part of our face came to exist in the first place. As with most of our sensory systems, we will have to travel back a few millennia to trace the origins of smell. While smelling under water might appear somewhat of an impossibility to us, the vast majority of biological life has lived in the seas. In those seas some 700 million years ago, there was a mutation in one aquatic vertebrate lifeform.[47] This lifeform still had neither eyes nor a brain, but some of their cells had developed the capacity to sense residual particles floating in the water surrounding it. This is essentially what a smell is: molecules left behind by some object or organism. These molecules might not be the same shape as the object itself; for instance, when you are smelling that rose, you are not necessarily detecting a specific "rose-molecule" but rather a set of particles that this certain plant happens to produce. Our closest non-vertebrate ancestor known to possess the sense of smell is thought to be the humble lancelet.[48] As stated before, this small creature developed a set of special cells some 700 million years ago, and these cells line their sides and passively pick up passing-by particles, allowing them to detect any aquatic presence without having to see or hear them. We often associate smell with a nose, but just as with hearing and vision, our sense of smell started with a mutation not yet associated with any particular organ. As we shall see later, this has given rise to quite a wide set of noses in the animal world. However, let us start on familiar ground and investigate the human solution to this olfactory puzzle.

Just like the rose may smell nice to attract pollinating bees, the first smellers of the sea most likely used their newfound sense in a similar way and by letting their existence be known to the outside world, they would increase their chances of finding a mate. However, *smells* did not evolve as a response to the *ability to* smell. Rather, the little aquatic amphioxi that suddenly could detect potential mates in their vicinity would have a better chance of spreading their new mutations on since this new trait would provide them with an evolutionary advantage and they would produce more offspring. Similarly, an organism that can detect its predator, or its prey, will be able to avoid some nasty surprises while also conjuring some of their own. Further down the line, animals would indeed also develop the ability to improve their natural scents, like the rose. Other less devoted smellers do the same but through artificial means, which is why the human perfume industry is a billion-dollar venture.

Sniffing for Smells

The olfactory system is unsurprisingly responsible for picking up and processing odour molecules and identifying their origins. Particles that are lucky enough to escape the nose hairs end up in the nasal cavity and at the top of this air filled cavern, past the winding nasal turbinates, exists a carpet-like surface just waiting for smells to stick to it in what is called the *regio olfactoria.*[1]

This human olfactory region is considerably smaller than in many other mammals. Not easily deterred by this setback, humans are generally pretty capable of detecting smell molecules with remarkable precision. While this might seem like unwarranted praise considering how inferior we are to animals such as dogs when it comes to olfactory prowess, we have specialised in detecting certain molecules. For some smells, we are able to detect one single molecule drifting amongst a billion others.[19] Potentially dangerous, putrid smells seem to be especially easy to register. You might again think this is unfair. After all, stinky odours like skunks or rotten food are often so intense that one might wonder why our noses needed to specialise in picking them up. In actuality, the relationship is the reverse — the reason we perceive them as so intense is because they signify something important, usually danger.

A smell is however often not due to just detection of a single molecule. Indeed, the odour detection threshold refers to the number of molecules needed to evoke a response. Similarly, the concentration of a smell at which it can be identified is called the *recognition threshold.*[19] This is all dictated by the way the smell molecule is built up as it needs to fit in with the different smell receptors in our nose. A rose does not produce one single type of molecule that it releases into the air but rather a whole set of them. So, when we detect a smell, it is most often a combination of molecules that have gotten stuck in the welcoming mat inside our nasal cavity.

The little hairs that pick up the smell molecules are doused with olfactory receptors. Since there are a great many smells, there obviously also needs to be a great number of smell receptors. Or rather, since we have so many different receptors, we are able to distinguish between all those smells. Evolution saw to this by having these smell-cells emerge from progenitors prone to mutations.[47] The genes coding for our smell-detectors are located in a position of the DNA that allows for several crossover mutations to take place; this means that the information telling a cell what it should be able to smell can flip over to another cell when they multiply, creating a new structure that might react differently to the smell molecule it was designed for if further mutations occur in subsequent generations. This makes for a substantial diversity within the human arsenal of olfactory receptors. As a point of comparison, the humble brainless lancelet possesses 40 distinct olfactory genes, whereas we humans have at least 339 DNA segments coding for smell receptors and 297 that in one way or another influence their development.[47,49]

Let us take a deeper look into how the presence of those smelly molecules is registered by our brains, and their signal relayed through the path outlined in Figure 3. Olfactory receptors are bipolar, just like the sense-relaying neurons in the retina[1]. This means that they have one *axon*, which is the main nerve fibre, and one *dendrite*, the branch that eagerly tries to pick up the sensory information they are specialised for. The fuzzy part of the olfactory region is made up of these dendrites. Each receptor has one single dendrite reaching into the nasal cavity. From this tendril, several little hair

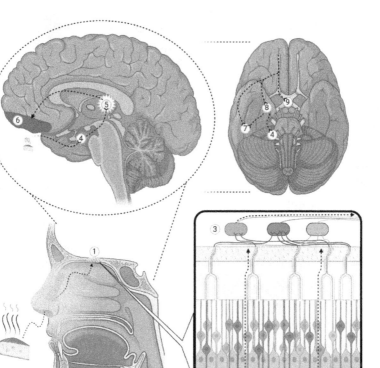

1. The first step for any scent is to stick to the olfactory epithelium at the top of the nasal cavity
2. A molecule binds to its designated receptor, creating a neural signal
3. Matching receptors signal to the same glomerulus, creating a specific signal pattern forwarded to the brain
4. The amygdala receives both direct input, which causes an immediate emotional response, as well as a copy from the cortex allowing a more contextualized interpretation of the smell
5. Having passed by the piriform cortex in which the amygdala sits, the scent reaches the thalamus
6. The orbitofrontal cortex allows us to perceive the smell
7. A direct projection to the hippocampus immediately elicits memories associated with a certain smell
8. The olfactory cortex represents all structures receiving direct input on a smell (1, 4 and 6), creating a complex and subconscious response
9. The hypothalamus is responsible for hormone production, and a smell can prepare it for an expected meal

Figure 3. How we smell. Starting from the bottom and moving upwards, you can trace the trajectory of a smell molecule and the subsequent neural impulse it causes. As you can see, the neural wiring allows smells to trigger immediate responses without us being conscious of them. Created using Biorender.

cells, or *cilia*, stick out covered in a thick fluid protecting them and making sure electric signals can be generated. When molecules happen to get stuck in this sticky fluid, they get dissolved and brought in for questioning. Should they fail to show proper ID identifying them as smell molecules, they will promptly attract security that will remove them from the premise in order to protect the sensitive hair cells. The smell particles will happily be let through and attach to their partnered receptor.[1] Simply put, certain molecules will exhibit certain features that allows them to match with other cells in the olfactory epithelium, allowing them to be registered as smells. The cilia on which the receptor cells are attached act like a net, letting the nerve cells cover a much greater area. Depending on how you are feeling, this process might not always be so easy however. For instance, when we catch a cold, more of the sticky fluid is produced, forcing the dissolved smell molecules to travel a bit longer. Many however just get tired along the way, get lost, and leave their partnered receptors, resulting in a dampened sense of smell during a cold.

Understanding a Smell

When the smell molecules find their target hair cells, this causes channels to open in an associated nerve. Ions of different charges then begin to flow through these new gates and create an action potential, or an electric charge, translating the smell particles' shape and structure into information comprehensible to the brain.[1] The pattern of this electrical signalling is essentially what makes up a smell, and while one type of olfactory receptor might fire in response to a certain molecule, most smells are built up of many different smell particles with different detection thresholds. A flower with a complex scent does not activate only one set of receptors, but will instead be playing the multifaceted clavier of your olfactory region.

Just like hearing the same note over a long period of time, if we are exposed to a background smell for a long period of time, the response to that specific molecule will weaken as you grow accustomed to a smell.

The unique electrical patterns generated in the nasal cavity will come together on the other side of the cribriform plate, the bone that separates the olfactory region from the brain. Here they converge in the olfactory bulb.[1] This acts as the first step in integrating the electrical information and sending it on to different areas of the brain. Unlike many sensory signals, smell will bypass the brain's distribution centre in the thalamus. Instead, it goes straight to the olfactory cortex. The signal also follows separate pathways to the hypothalamus, which is part of the limbic system that controls behaviour and emotions, as well as the amygdala where our fight-or-flight responses are governed. Both the hypothalamus and the amygdala represent basic brain functions essential for survival. Similarly, the olfactory cortex is believed to be the oldest of the cortical regions.[50] Smell, therefore, seems to be hardwired in such a way that it allows us speedy and reflexive responses. Considering animals developed smell before we developed a brain, our noses may have been the most crucial sense for our early survival.

Before the signal is sent off to its target brain areas, the olfactory bulb makes sure to collect and categorise the incoming electric signals.[1] Essentially, it is in the bulb that the nature of a smell is decided through a primordial archiving system. The axons of the olfactory receptors act as highways into the bulb, clinging to the underbelly of the brain. Here, they gather together in spheres that are called *glomeruli*, resembling tiny balls of yarn. These little balls receive input from specific olfactory receptors in the nasal cavity. Naturally, this means that many different glomeruli will activate in response to a smell as they are all built up of multiple molecules hitting different receptors. In true democratic spirit, the ones that light up the brightest decide what smell is to be relayed; so, if you smell both bread and freshly made coffee, chances are the glomeruli detecting coffee in your olfactory bulb will have the upper hand as coffee has a lower detection threshold. A primary dendrite belonging to a *mitral cell then* reaches into the glomeruli. Through this neural offshoot, the electrical impulse is then relayed through the mitral cells into the brain areas we previously mentioned: the olfactory cortex, hypothalamus and amygdala.[1]

Thinking With Your Nose

For all the benefit the sense of smell must have given our ancestors, it has a strange quirk: when asleep, the human olfactory system essentially shuts down.[51] One can speculate that this allows us better rest through sparing us from being awoken by any olfactory stimulation, but it certainly comes with its fair share of risks; for instance, the otherwise rather recognisable odour of burning wood could fill the air without our being able to notice it. This has in fact been tested by having sleeping subjects smell pyridine, a deadly smell associated with fires.[51] No matter how deeply a participant was sleeping, the life-threatening odour was incapable of waking anyone. In this case, I think most of us can agree that investing in an artificial nose dedicated to smelling smoke is a valuable investment, though luckily that is exactly what a fire-alarm is.

　　When awake, a dangerous smell will instead have a very clear effect in signalling that we are close to something we should be avoiding. It was generally believed that humans could detect roughly 10,000 different scents, but recent studies completely shattered that estimation and instead argue that humans can differentiate between 1 trillion different smells! That is 1,000,000,000,000,000,000 different ways the olfactory receptors can be combined in their firing patterns to various smell molecule combinations. This feat however gets harder to accomplish with age; people between the ages of 20–40 will correctly identify 50–75% of a scent, but this dwindles to 30–45% when aged 50–70.[19]

The instinctive aversion we feel to noxious smells is also com-bined with other physiological responses. For example, the trigeminal nerve can be triggered by irritant particles.[19] This nerve stretches across your face and can send strong signals of pain when triggered. So not only will your brain tell you that you are at risk of being injured by whatever is giving off that horrible smell, it may also try to warn you from approaching it by making you experience physical pain. Like the pain telling you to remove your hand from a scorching metal surface, this pain aims to force you to move some-where safer. Furthermore, your body will react by trying to remove further sensory insults. For instance, being exposed to a pungent smell might cause your eyes to tear and your nose to run, which may make it harder for you to breathe. The increase in fluid production in your eyes and nose is to dilute whatever toxic molecules you were unlucky enough to breathe in,[19] while the shortness of breath may be caused by your windpipes constricting in order to stop you from breathing further particles.

Despite the olfactory system appearing to be relatively conserved between species, there certainly seem to be quite substantial varia-tions even within our own species. For instance, sommeliers make a living based on their keen sense of smell. One can speculate that the olfactory receptors' inclination for mutation, which allowed our olfactory range to evolve, is responsible for this diversity. Having an overly keen sense of smell is not necessarily a benefit, however. It is not a sense that you can simply turn off. We might not think of olfac-tory disorders very often, but for some people their sense of smell can cause severe handicaps.

Hyperosmia is a condition caused by a lower detection threshold for odours. While it is hard to know what causes this disorder, cer-tain genetic variations have been identified to cause hypersensitivity to certain smells.[52] It is unclear whether there is a single genetic mutation that causes general hyperosmia or if several mutations for different smells are linked to the phenomenon. There also seems to exist an association with a general sensitivity to chemicals, which can often be seen during natural hormonal cycles which interfere with chemical signalling.[53] The cause may also be more central within the brain.[54] Interestingly, it is quite common that patients suffering from migraines express heightened sensitivity to smells as well as light or loud noises, and patients with epilepsy may also describe this heightened olfactory awareness in relation with seizures, with some experiencing certain smells prior to onset of an epileptic fit. The latter concept is similar to the popular myth that a stroke is accompanied with the smell of burnt toast. The myth regarding the smell of burnt toast accompanying a stroke comes from the sensation patients might have had when the olfactory cortex, or the olfactory pathway, was affected. In stroke, any signal being processed in the region which is deprived of blood flow, and therefore oxygen, is likely to be distorted. Since this can happen in any blood vessel, a stroke in the olfactory bulb could theoretically lead to a patient experiencing a smell which is not physically present. Such olfactory hallucinations are common in both stroke and epilepsy. Both stroke victims and patients with epilepsy might perceive the smell of burnt toast, but this is because of damage to neurons in the olfactory cortex. However, as these two disorders can affect any region of the brain, it is by no means certain that these symptoms will manifest in all patients with stroke or epilepsy.

Olfactory hallucinations of this kind are called *phantosmia*.[55] Much like phantom pain experiences from the nerves innervating a long-lost limb, this involves sensing something that does not exist, in this case a smell. When this happens, something has gone wrong in either the brain or the wiring that connects it to your nose. Patients suffering from migraines or epilepsy can get these hallucinations due to sudden changes in electrical signalling brought on by pathologies that still are quite poorly understood. Similarly, patients who have had extensive surgery in their nasal cavity might experience phantosmia due to damaged olfactory receptors.[55] While some sedatives might alleviate symptoms, the most effective solution to this specific problem is ironically to surgically remove the remaining receptors as well.[56] This obviously cures the problems caused by the faulty spark plugs that make up the olfactory region but it also eliminates any sense of smell. For many of us, this would be an unthinkable solution, causing an even bigger problem than it solves. We should remember that these patients have no way of separating real scents from hallucinations, so if they were to later sustain damage to their olfactory cortex, any smells perceived would seem very real to them and if it's an unpleasant smell, their receptors won't be able to habituate to the smell as the smell does not really exist. Hallucinated toxic smells will still cause your eyes to water and your lungs to contract. You will never get used to it since the receptors cannot be habituated to something that does not really exist. Cases like these serve to remind us how deeply integrated we are with our senses, and how crucially dependent on them we are.

While some smell things that are not there, others suffer from the inverse. The absence of an ability to smell is called *anosmia* and can happen due to a series of different disorders, either affecting the nasal cavity itself, like polyps or viral infections, or the neural network connecting the nose to the brain, often through trauma or ischemia.[57] On the other hand, the unfortunate state of misidentifying smells is called *parosmia*.[57] Patients with parosmia often misinterpret harmless scents as highly unpleasant.[58] The smell of grass after rain might instead be experienced as a burning cow pasture next to a chemical processing plant. Much like olfactory hallucinations, this can naturally be caused by central damage to the brain. More often than not, the cause is further in the periphery however. It would seem as if upper respiratory tract infections can inflict serious damage to the olfactory receptors when stampeding through the nose and sinuses. A damaged neuron here will have severe repercussions further up the line, with the faulty electrical signal reaching the olfactory bulb and lighting up a completely different set of spherical glomeruli. The brain can only discern this last step, causing the smelly misunderstanding. You might wonder why this would automatically mean that the smell is perceived as something unpleasant. If the wires have just gotten crossed, is it not just as likely that you would interpret the smell as something sweet or inviting? Surely that would be advantageous to our well-being, but consider the reaction of the olfactory bulb when it gets the faulty signal. Here, the smell is experienced as something completely new, something strange and unexpected. Generally, when we are faced with something unforeseen, the safest course of action is to take a step back and avoid it until the brain has reached a consensus on the nature of that smell. Misinterpreting a scent as something negative might therefore be a survival mechanism, though it should be stated that in very rare cases parosmia has been experienced as pleasant odours as well.[59]

Guided By Our Noses

At the beginning of this chapter, we brought up the notion that smell is such a human experience that robots would be unable to reproduce the sensation. We also stated that this is wrong. Smell is nothing more than the interpretation of small particles produced by any given object. Any mechanism capable of identifying these particles is for all intents and purposes capable of smelling. An electronic nose will of course not have any intrinsic pleasure or aversion related to the process however; it might say that it detects acetone by a quantity of 120 parts per million but it will not try to roll off the table as a means of self-preservation.

Is such a concept completely incomprehensible though? When we are faced with a new smell, we immediately draw conclusions based on what we think we already know. A sweet smell is probably not so dangerous, whereas something unpleasant should be avoided. When we experience pleasure or disgust to a scent, it is in relation to the pre-programmed responses that evolution has installed in our brain. For example, an animal that runs away from a fire before the flames touch it will most likely live to produce more offspring. By not having to re-learn this every generation, we increased our chances of survival, much like the bee will instinctively know that a particular smell will lead it to the sweet nectar it seeks. In that way, there is no universal truth on how a smell should be interpreted. It is all a form of biological programming. In theory, it could be possible to program robots to react to certain smells the same way we do. Like evolution took care of our programming, we will in turn decide how the future of artificial intelligence should react to the same parameters. Evolution might have gone digital, but that does not mean it is going to stop — on the contrary, it would seem to be increasing exponentially with smarter and smarter devices.

Despite the challenges of identifying single smells, we have come quite far in artificial olfactory technology. The electronic nose works according to the same basic principles as our own olfactory system, which might not come as a surprise as we were its designers. While different forms and shapes of this technology exist, we will focus on their shared mechanisms. Keep in mind that an electronic nose is not a physical robotic nose but a machine that is able to pick particles out of the air. It singles out molecules, compares them to its database of known substances, and counts the number of molecules present.[60] The three stages in this process resemble the steps in the human olfactory network: we have a delivery system, a detection system and a computing system. In humans, this would correspond to the olfactory receptors, the olfactory bulb, and the olfactory cortex. Essentially, both forms of smell detection transform chemical molecules into electrical signals that are in turn processed. The human olfactory receptors have developed to respond to a specific type of molecule. This would be a very inefficient way of designing a robot. Ideally, an artificial sensor would detect all airborne molecules, but react differently depending on their properties. There are however combinations of electronic and organic components, producing bio-electronic noses which make use of proteins that have been cloned from organic tissue. Such bio-machines can mimic human perception quite well depending on the original proteins, and with tissue from a human host they perform with impressive precision.[60]

The use of robotics to help us see things in the distance, or hear things imperceptible to the human ear, might have their obvious advantages. What would we really stand to gain from an enhanced sense of smell? Today we famously use dogs, whose sense of smell is far superior to ours, when aiming to find things or people that we want to locate. This sometimes involves very dangerous circumstances ranging from war zones to dealing with lethal chemicals. A smelling machine would work towards minimising the risks faced by these animals, while also adding further sensitivity. We also make use of different smells as a way of identifying danger. Gas used in homes or businesses is by law required to have a strong smell added since it is naturally odourless. The added odour helps us identify leaks before the situation becomes critical and devices that measure smells in order to ensure they are readily recognisable by humans are called *olfactometers*.

Relatively new insight into the impressive sensitivity of smell has revealed a completely new potential field of use. Dogs are now being trained to sniff out Parkinson's disease in patients.[61] Some evidence even suggests that this method would allow us to identify the disorder even before it is detectable by other tools presently at our disposal. Surprisingly, this study is based on the remarkable talent of Joy Milne.[62] Under controlled experimental conditions, this Scottish lady was able to correctly identify six patients with Parkinson's disease out of a line-up of 12. She did however also pick out one of the healthy controls alongside the patients. What makes the story even more remarkable is that she also developed the disease shortly afterwards. Inspired by Joy's unique but useful mutation, the medical field is expanding its arsenal by bringing noses into the mix of diagnostic tools. If using odours in the medical field sounds strange, remember that before we were able to reliably measure blood sugar levels, a universally used technique was to smell a patient's breath. A sweet smell would reveal ongoing ketosis, meaning that the body is breaking down its fat deposits due to a lack of sugar. Milne's story sounds like something borne out of a fairy tale or urban legend, but the results are nevertheless there, reminding us how amazing the biological world can be. For those of us without Joy's olfactory talents, despite all our knowledge and technological advancements, we still cannot see further than our own noses. We still have a long way to go, which I find both humbling and reassuring.

Smells can obviously reveal hidden truths about what is going on in the body. In return, the body will respond with different physiological expressions to certain smells. While a toxic smell will make someone averse to ingesting any more, pleasant scents also have important biological functions. When you feel hungry and you smell something enticing, your stomach often begins to rumble. The reason for this unfair biological blackmailing is that your body is preparing for the anticipated food. Your sense of smells invites the first step of digestion, the *cephalic phase*.[1] We will not go too much into the specifics of the gastrointestinal tract, but the concept serves as a wonderful example of how the senses control your body without our interference. As the molecules released by the food attach to the surface in our nasal cavity, they activate the olfactory receptors which form electrical impulses that reach the brain via the bulb, and the appetite centres of the amygdala and hypothalamus are then activated. The vagus nerve then carries information to the heart, lungs, and the digestive system.[1] This nerve also tells the stomach to start preparing for business. The acid *gastrin* will start being released into the hitherto empty pouch long before any food gets there. It will also tell the stomach to get a move on, as the kneading of the gastric walls helps dissolve the food into the tiny building blocks of amino acids, minerals and vitamins that our body needs. When you are hungry and yet to eat, this kneading in your stomach causes the rumblings that reveal your culinary predicament.

When we touch on how smell is used by some other members of the animal kingdom in a later chapter, we will briefly explore the concept of pheromones. These smell molecules are airborne hormones that, when smelled, have a biological impact on another member of that species. It has been widely debated whether humans are somehow communicating through this smell-mediated mechanism or not. However, humans lack the *vomeronasal* organ that makes this form of communication possible in the first place. By contrast, the animal kingdom is full of fascinating examples to pick from and we will explore some of these later.

Regardless of the messages conveyed by your eyes and ears, the tiny particles that rush into your nose as you breathe in also speak a language of their own. It may not be as clear of a communication platform as the pheromones of the animal world, and we do not use it very much to engage in debate or discussion, but before language allowed us to communicate a danger we could not yet hear or see, we had to rely on our other senses. Even today, regardless of whether you are exposed to a cacophony of indistinct voices or enjoying complete silence, your nose and the smells it deciphers will be able to imbue further meaning to your environment.

Chapter 5

Taste

A Taste for Flavour

Most of us can probably agree that a sense of taste adds significant value to our everyday lives. Few feelings compare to the joy of finding a forgotten treat in your pockets, and a long day at work might be held together by the promise of a favourite meal later that evening. We take great pride in the way we prepare our food, making it healthy and primarily tasty. Our human taste buds hold many of us in a vice-like grip and our continuous struggle to appease these sensory overlords makes up a cornerstone of the human psyche. Civilisations might have looked very different without our perpetual aim to satisfy our taste buds. What would human history have looked like without the spice trade, for example? Would the trading network enabling the Industrial Revolution have been put in place if the Brits' desire for tea had not prompted them to conquer a quarter of the entire world? As a side note, we should briefly touch upon another aspect of smell which we did not explore in the previous chapter, retronasal olfaction. In reality, tea or spices have little to do with actual taste, and more to do with olfactory receptors in the back of your throat picking up and interpreting odour molecules, with this contributing to the perception of a substance's flavour. This happens more centrally in the brain, as we will see later, but since these foodstuffs rely more heavily on olfaction rather than taste as detected through the tongue, it is important to make this distinction early on.

Still, there is no denying that taste has played a vital role throughout human evolution. As this sense went from serving as a tool for survival to a device for luxury, the direction of human society shifted. Rather than propelling us towards survival, an excessive adherence to this primordial instinct, albeit with a focus more on flavour than taste, causes more and more deaths every year by leading to obesity, not to mention the role it plays on the world's uneven distribution of resources. The cultural impact of food, however, is not the topic of this chapter, and it is better left to be explained by more expert sociologists and historians. We will, however, take a deeper look into how the double-edged sword of taste came to be.

Taste, the ability to distinguish between sweet, salty, sour, bitter and umami, might seem like an obvious contender for a placement in the pantheon of human senses. In reality, taste was initially somewhat of the sensory stepchild for several 100 years, not being considered a proper sense until around 150 BC as reflected in the Roman author Aulus Gellius' definition of the five senses.[2] Rather than being a sensory system on its own, Aristotle considered taste to be a sub-specialisation of *touch*, meaning that taste got to piggyback on the sensorimotor system for about 200 years. If this seems a bit counter-intuitive, consider that many other types of sensory inputs are still attributed to touch. Pain, for example, is relayed through a completely different pathway up the spinal cord than blunt pressure. The ability to distinguish between hot and cold is built up in a similar way, physiologically separate from what we normally would refer to as touch. This is why certain spinal cord injuries will lead to the inability to feel pain on one side of the body while the sense of touch is instead affected on the other, as we shall see in a later chapter. I am not necessarily arguing for inaugurating pain and temperature as senses, but I think the above very aptly illustrates how culturally biased our approach towards the senses is. For now, let us be grateful to Aulus Gellius for recognising taste as a sense and explore it further.

Let us go back in time about 250 million years. The first mammal, a shrew-like creature, scurries along the dense underbrush of the late Triassic period.[63] Dinosaurs will not be the dominant life form for another 50 million years, and for now the evolutionary playing field is wide open. Our mammalian predecessor knows well how to forage for food, but as it spreads across new lands it encounters strange new berries in the arid inlands of Pangea. It gives it a quick nibble. Bitter. The shrew-like being continues its search. Eventually it will find something more palatable, a sweet fruit lying by a tree in a desert oasis. Satisfied, our ancient forefather enjoys a well-earned respite. A fellow member of its species will shortly make the same trip. Having been injured in an encounter with an unfriendly reptilian, they are a bit slower. It tastes the same berry as its friend. Nothing. The damage from the fight has rendered it both slow and incapable of taste. Unburdened by this fact, it gobbles up the pulpy content which, unknown to him, is lethal. Half an hour later the animal lies dead by the berry bushes, serving as a visual warning to those who would come after.

Taste has been vital in keeping animal life geared towards survival throughout existence. The reason we seek things that are sweet is because it reveals a content rich in carbohydrates, highly efficient in producing energy.[64] The bitter taste of a leaf, by contrast, hints at a lack of valuable nutrients and instead exposes an object as potentially poisonous to us. So does that mean that animals have different sensations to taste depending on their dietary needs? Pretty much. Cats, for example, lack the ability to taste sweetness altogether, as the genetic material coding for the appropriate sweet-taste receptor has simply been turned off in cats through their evolution over the last millennia, making them seek out fat and protein-rich diets instead.[64] Other animals also show similar deviations from our human-centric taste-buds, as we will see in a later chapter.

Taste and the Need to Feed

The importance of taste for survival has diminished drastically since humans evolved and climbed down from the trees. The consequences of eating certain foods have been very well mapped, and it is exceedingly rare to find a foreign substance in the wild which has not already been indexed in terms of edibility. This is the primary reason why we no longer shun the taste of bitterness despite most animals avoiding it. Ironically, bitterness has instead become somewhat of an acquired taste and is even advertised in beers and other beverages. Our primitive beginnings still make themselves known however. Two specific stages of development see us revert back to our taste-oriented ancestors: as young children and during pregnancy.[65]

It is a truth universally acknowledged that children love sugar. Though it may be fairer to say that everyone loves sugar, but children are generally more vocal in seeking it out. From an evolutionary perspective, this makes a lot of sense as vegetables are generally low in sugar and therefore contain few calories. Calories are, of course, vital to survival, and even as adults we might find it hard to push mind over matter and tell ourselves that we could do without that extra chocolate bar when shopping for food on an empty stomach. The survival instinct also returns during bouts of pregnancy. It takes a lot of energy to build a new human, so logically this makes the mother-to-be keener on consuming foods rich in nutrients. It also means that she needs to avoid toxins, which manifests as an increased aversion to all things bitter.[65]

So far we have dealt with sweet and bitter. Before we delve deeper into the physiological mechanisms, it feels apt to touch on the purpose of the other tastes. After all, we enjoy sweet things because we need them, we avoid bitter things because they harm us, so why are we capable of identifying salt, sour and umami? Salt might appear to be rather obvious. We need salt to survive. Nothing really tastes like salt except for the actual mineral itself, meaning that if we taste it, we get it. The reason for this is to keep a level fluid balance, or homeostasis, in the body. As we lose or gain fluids, we need to alter our salt levels to balance it. After a long hot summer's day, you may have found yourself rummaging through the pantry for a bag of salted nuts or crisps. This "salt-hunger" is your sweaty body's way of letting you know that it has lost more salt than water through sweating, and asking you to remedy that.

Sourness is a more interesting taste. We generally tend to avoid it, and many of us can perhaps recall that first time we bit into a lemon. Yet, we still seek it out. Fermented, sour, foods are popular amongst many cultures and I personally grew up eating the perennial serving of fermented herring, which I have since learned is considered a biological hazard by airlines. This acidic taste is also prevalent in artificial sweets, beverages and foods, and that can certainly be no coincidence. But what is the purpose of such a distinct taste? The reigning theory is that our detection of sour would have enabled our ancestors to identify vitamin C, which is essential for the Great Ape family of primates to which we belong.[65] The lack of this vitamin has caused countless deaths at sea through a disease called scurvy before ships started bringing lemons with them on long voyages. Vitamin C is needed for the body to produce collagen, which is the primary building block of most anatomical structures. Without it, wounds cannot heal, teeth may start to fall out, and the blood vessels will start to leak, which is exactly what happened to the scurvy-ridden sailors of yore. Cats, once again the prime example of sensory deviants, have no need for exogenous vitamin C as they produce it themselves in the liver. They can, however, still taste sour foods, most likely because it reveals rancid meat. This may well also be the reason for our own human cognitive dissonance: we need to seek out the vital sourness of vitamin C, but must avoid meat that has gone bad.

Umami brings back some logic into the mix, representing the taste of cooked meats.[65] What makes less sense is that it was not until 1908 that a Japanese scientist, Kikunae Ikeda, isolated the taste by proving that it was caused by the amino acid glutamate.[66] The introduction of meat into our diet is relatively new compared to fruits, so it does make sense that we are less receptive to its taste but one might find it odd that it is only in modern times that we recognised the concept, with the scientific community not accepting the term umami until 1985. The real cause might be that umami is not really a well-defined taste on its own, but is instead brought out when combined with other tastes and smells. In isolation, it is essentially only during fermentation that glutamate is made available in meats.[65]

On the Tip of the Tongue

The sense of taste is detected, to no-one's surprise, on the tongue.[1] At least primarily, but there are also taste receptors in the soft palate, the upper part of the oesophagus, as well as in the pharynx; you can essentially taste with everything from your tongue to your throat. Consequently, the identification of taste is a lot more developed than one might first expect. While flavours rely on a wide set of detectors across the mouth and throat, what we usually refer to as taste can only be really experienced on the tongue where it is picked up by so-called papillae, which are small rounded bumps. These bumps are surrounded by a small indentation in the tongue, which allows the taste molecules to concentrate around them in order to be picked up by receptors on their walls. The pathway of these molecules and the signals they produce are outlined in Figure 4.

The nature of these taste buds has recently been the topic of much debate. Historically, it was believed that different types of taste receptors were only capable of detecting a certain taste. A taste-belt of this kind suggested that sweet was only perceived at the tip of the tongue, while bitter foods were detected further back. In truth however, all taste buds seem to react to each taste. The reason for the initial misunderstanding was that all taste buds have different thresholds for sweet, sour, salt and umami, allowing them a robust specialisation.[67] This tasty specialisation goes something like this:

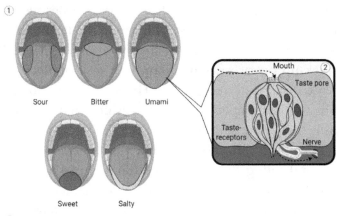

(1) The general distribution of taste buds sensitive to the five tastes

(2) A taste bud, consisting of the taste pore trapping a molecule before it is registered by the bud's receptors and transformed into a neural signal

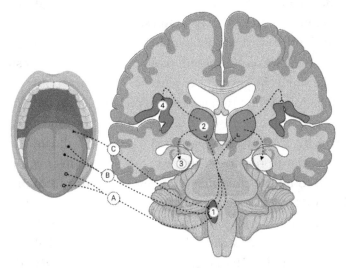

(1) The medulla oblongata hosts a set of nuclei revelant to taste, influenced by three nerves: the facial nerve (A), the glossopharyngeal nerve (B), and the vagus nerve (C)

(2) The thalamus redirects the incoming signal to relevant brain areas

(3) The amygdala creates an instinctive emotional response that doesn't rely on a conscious sensation

(4) Tastes are perceived in the gustatory cortex

Figure 4. **The path of taste.** The top image illustrates how taste buds are concentrated on the tongue and how they look. While these regions are particularly perceptive to these tastes, we now know that taste buds are spread across the entire tongue. The bottom schematic shows how the three nerves conveying taste information projects to the brainstem, where the electric impulses are conveyed to different parts of the brain. Created using Biorender.

there are three distinct types of papillae located on various parts of the tongue and all have different sensitivities to the tastes. At the tip of the tongue, we find the fungus-shaped *papilla fungiforme*. These primarily pick up on sweet, umami and salt taste molecules, and make up 25% of all taste receptors.[19] The sides of the tongue are inhabited by the leaf-shaped *papilla foliate* which make up another 25% of taste buds and address our ambivalent relationship with sour. The remaining 50% are the crater-like *papilla circumvallate* which are found lining the base of the tongue, picking up on all things bitter. These distinct areas are further cemented by their isolated innervation, with one facial nerve being dedicated to each type of taste bud.

Thus far we have been talking about *taste-particles*, which in turn give a distinct sensation when picked up by the tongue's receptors. These particles naturally have names. Unlike smells, tastes are a bit more straightforward in their impact on the brain, being caused by specific molecules as compared to olfaction's wide mix of substances. Much like smells however, taste particles have threshold values which need to be met if the brain is going to pick up on them. While several substances can cause distinct tastes, there are some that are more prevalent than others. For simplicity's sake, we will run through the most important ones for each basic taste. Sweetness is for example perceived to be the presence of *sucrose* at a concentration of 20 millimoles (mM).[19] If you are wondering how much that would be in terms of amount, it is exactly 0.24 grams of carbon-12 (12 grams of carbon-12 acts as the standard for one mole unit). Salt and umami are both detected at 10 mM each, and comprise of *salt* and *glutamate*, respectively. Citric acid is what gives food a sour taste, and with a detection threshold of 2 mM, probably very few of us remember our first encounter with a lime very fondly. The true enemy is however quinine. This refined tree bark has saved thousands of lives through its use as a malaria treatment, considered by the World Health Organisation as one of the most effective and safest medications in the world. Our less developed ancestors would probably never accept such a statement — quinine is horrifyingly bitter, and with a detection threshold of only 10 μM, corresponding to a flake the size of 0.00012 grams of carbon, not much would be needed to send early man running for cover.

On average, our tongue is home to about 4,000 taste buds capable of detecting substances through their taste receptors.[19] As the taste buds reside in little craters, these essentially make up tiny little tasting pools spread around your mouth. Small taste pores help our taste receptors pick up on the chemical signatures floating around in these little pools of sweet, sour, bitter, salt, and umami. Since they are almost constantly surrounded by taste-particles, this is quite a tiring ordeal for our sensitive receptors. As a result, the life of a taste receptor is short but sweet, or any other sensation depending on its location. After 2 weeks of work, the receptor will die off. New additions are produced by basal cells located deeper within the taste bud. This is perfectly logical considering the high risk these cells have for encountering toxic substances that could damage their function. Therefore, having a rapid life-cycle means we are not left without a sense of taste for long should we accidentally put something dangerous into our mouths.

Right in the middle of the taste pore we once again find microvilli.[19] These elongated cells are extensions of the underlying membrane, and carry on their stem the receptor proteins that pick up the chemical compounds associated with each taste, with there being one type of receptor for each of the five basic tastes. As we will discover, microvilli are a recurring element in sensory neuroscience, and while sometimes referred to as hair cells as their appearance certainly makes them appear as such, they should not be confused with the hair we have on our skin; the former is an elongation of a membrane while the latter is made of keratin, but both are involved in sensory processing as we shall discover later when dealing with touch.

At this point you could make an argument not only for taste being a specialised form of touch, but for the individual tastes being considered distinct senses of their own. The physiological method of picking up the taste particles differs greatly between them. Salt and sour substances open up a specific type of channel in the microvilli, activating sodium channels as a response to salt, while instead preferring hydrogen channels for acids. Salt is after all usually from sodium-chloride and acid compounds are rich in hydrogen. Sweet, bitter and umami instead activate another type of channel protein which causes an influx of calcium into the cells. As with all other senses, the consequential movement of charged particles in and out of the cell causes the build-up of an electric charge, which in this case is sent on its way towards the brain through synapses at the back of each taste bud.

What Does It Taste Like?

In the relay of gustatory information from the mouth to the brain, not only do we have three different ways of translating chemical properties into neural impulses, but the route they take to the brain is also completely different.[19] What makes it somewhat convoluted is that the neural innervation does not match the way the different channels are divided. As you can see in Figure 4, although the different tastes can be detected across the entire surface of the tongue, they are concentrated in different regions. Tastes detected on the front of the tongue are carried onwards by the facial nerve, more specifically its chorda tympani branch, while the glossopharyngeal nerve innervates the rear. Finally, the vagus nerve picks up those taste molecules that are registered even further back in the throat. All of these pathways are cranial nerves carrying sensory information straight to the brain. Irrespective of the sensory pathway used, the end station of this express highway is nevertheless the same: the *nucleus tractus solitarius* located in the brainstem.

This part of the brain acts as a great regulator of the autonomic nervous system, modifying the automated background functions such as breathing. Here, the tastes are segregated, resulting in separate taste clusters that give us the five basic tastes.[19] From here, the information is forwarded to the amygdala, our alarm central, which tells us whether we should be satisfied with the contents of whatever substance we just put into our mouth, or if the restaurant bill will include a trip to the emergency room. The solitary nucleus also sends the signal onwards to the thalamus, where essentially all neural signals are processed for future reference. From here, the sweet, sour, salt, bitter and umami signals are sent to the gustatory cortex where we are finally able to cognitively make sense of each taste. Similar to smell, this means that we can react to something we have tasted even if we do not necessarily know what it is, as the amygdala works on a different branch of the network.

The perception of what something tastes like or what flavour it has does not only rest on the sense of taste, but also smell and touch. If you were to look at a human face in profile, and peel off a little bit of skin and muscles, you would find the trigeminal nerve stretch across the face like a face-hugging crab. It has three main branches, as its name suggests: the ophthalmic, surrounding the eyes, maxillary, reaching towards the nose, and mandibular, grabbing the chin, mouth and cheeks.[19] This widespread net is what makes it possible for light, smells, and foods to cause actual pain as smaller branches from the three main trunks also contribute to pain perception.

This twisted root network hidden in your face is responsible for giving food its texture, relayed through the mandibular branch. It is also responsible for producing the concept of *spiciness*, which essentially can be called a sub-specialisation of temperature control, with this being in turn a form of touch or somato-sensation. *Capsaicin*, the active ingredient in peppers, notably confuses the body's nociceptors, which mediate pain and cause the brain to also register an increased temperature in that area.[19] This is why you do not want to get spices in your eyes, and if it is strong enough capsaicin might even hurt the skin itself. Depending on your background, you might have found yourself sweating after having visited an authentic Indian restaurant. This is because of the effect capsaicin has on your temperature receptors, tricking your brain into thinking that you have been teleported to a tropical island. It is no coincidence that spices are a key ingredient in many recipes. The local increase in temperature, at least as your brain perceives it, leads to increased salivation which facilitates eating coarse food, such as unprocessed rice[68]. It also makes the brain produce endorphins, which is the main organic pain killer released into the bloodstream in response to physical harm. Since we are not in any real danger, we benefit from the endorphins without having suffered the negatives, deriving a sense of well-being despite the light gustatory masochism.

However, the effect of a spice, or any other taste for that matter, might not necessarily be the same between individuals. It is believed that the perception of salt, for example, decreases with age.[19] This is primarily due to the loss of those taste-sensitive papillae we talked about before, which decrease in concentration after the age of fifty. Salivation also generally decreases with age, meaning there is less solvent that can bring each taste particle into the little craters and pores where the receptors can connect with them. As a result, older age often brings with it an increased consumption of salt, not because the body needs more salt, but because the brain cannot register its consumption. However, with more salt in the bloodstream there must be additional water if the body is to maintain homeostasis, resulting in a general increase in blood pressure. While this is true for everyone who consumes a lot of salt, the effects are particularly dangerous to the elderly, many of whom may have co-existing cardiovascular disorders.

Flavour — A Convergence of the Senses

Now, armed with the senses of taste, smell and touch, we can construct a proper flavour-profile. When your waiter interrupts you in the middle of dinner asking how your food tastes, it is more specifically the flavour they are asking about, as cooking is much more holistic an endeavour than simply throwing chemicals at the taste buds. You might have heard that if you pinch your nose while eating, the food will taste very different but that is not quite true. The food tastes the same, but the flavour is diminished due to the lack of complementary olfaction. This highlights that what we colloquially refer to as taste should actually be referred to as flavour, as it is really a combination of three separate senses: *taste, smell* and *touch*.

Apart from letting us detect poisonous substances and allowing the nuances of flavour, our sense of taste also takes part in a multisensory signalling cascade. We have already seen how visual and olfactory encounters with food send a signal to the stomach, allowing it a head start on the digestion that is to come. This happens during the *cephalic stage* of the digestion process, referring to this process happening all in your head; there is a promise of food, but this is yet to be delivered.[69] The pact is finally sealed when food enters the mouth, bringing some aspects of flavour into the mix. The food is crushed in the process of mastication, or chewing and enzymes in the saliva immediately start breaking down carbohydrates. Some other interesting things are also happening at this point. For example, you will continue chewing. Just like breathing, it is not something you actively think about, even though you have the ability to control the rate at which it is carried out.[69] This is where touch comes into play again. Nerve endings in the teeth and jaw are sending information to the brain to engage the chewing motor program, and depending on the food, that signal will regulate the strength and frequency of the chewing. It also automatically closes off your windpipe to make sure you survive the endeavour.

It might sound somewhat counterintuitive, but the taste of the food has very little to do with the actual digestion. One could argue that this makes sense considering that the food ends up in the same place anyway, regardless of its chemical composition. Sure, something sweet could warrant an increase in enzymes targeting carbohydrates, while chewing on a block of salt would employ a completely different chain of events. This happens in the stomach however, which is arguably much better placed to make those calls.

If the stomach is the CEO, then the tongue can be considered the concierge. If the stomach were to produce specific enzymes every time we put something in our mouth however, using taste as a toxin-detector would become redundant. Therefore, when we decide we like the taste of something, we voluntarily swallow the substance and in turn pass on responsibility to the stomach. By contrast, when we spit out those bitter fruits, we spare our gastric organs some work. Kneading food does consume energy after all, and consuming energy without being sure of receiving any would not be an emblem of a well-designed system.

Fortunately however, evolution does not produce flaws of that magnitude. Everything our body is capable of has developed for one very specific purpose: survival. Humans not having wings might feel like a bit of an oversight, but evolution saw it fit to give us denser, heavier, bones instead. This sense of purposeful design is also seen in how taste is dealt with and perceived across the animal kingdom. For instance, while cats do not taste sweetness, one may argue that this is a major benefit to feline life. They are carnivores and so do not use carbohydrates in their energy production, relying instead on proteins and fat in a species-wide ketogenic diet. As a result, they will never hunt for sweets. Sugar, as most of you know, is one of the most addictive and dangerous substances, killing countless people each year. While curiosity might have killed the cat, sugar at least certainly did not. Much like smell, taste acts as another wonderful example of how different our perception of the world can be between individuals and between species.

Who Decides What Something Tastes Like?

I believe taste is well suited to illustrate some of the problems that arise when we aim to categorise a sense, and how the context of our own lives influence the way we think of perception. As illustrated with the cat who is bereft of all things sweet, many animals do not perceive the same five basic tastes as us. Does that mean there are tastes we cannot comprehend? Is there a gustatory version of infra-red, a taste as unattainable to us as sweetness is for cats? Much like the senses themselves, there is no true definition of what constitutes a basic taste. The criterion for being able to classify something as a taste is if there is a receptor for a certain chemical structure. This however ignores examples like the hummingbird's affinity for sweet-ness despite it lacking the dedicated physiology to detect it, which we shall go further into in a later chapter. Perhaps we should look into different brain regions instead and see which areas light up when we put something into our mouths. Still, I think most of us can agree that experiencing food goes beyond purely activating differ-ent sets of on-off switches on the tongue. Renaming taste to something as technical as "peripheral chemoreception" would argu-ably tarnish any dinner date.

Perhaps trying to appeal to all organic life by using the same syntax for defining taste would do us all a disservice. In the end, it seems likely that our basic tastes, much like our senses, will continue to be subject to cultural interpretation. Culture, however, is a fluid concept that never stops evolving. Just because umami was accepted in the 20th century does not mean that we have filled our quota of acceptable basic tastes. There is a lot of effort going into categoris-ing the chemical and tasty properties our bodies are able to pick up on, so we will finish this chapter with a short rundown of these potential future taste candidates.

Much like we need salt for balancing our fluid levels through homeostasis, our muscles are reliant on an all-important mineral that we cannot produce ourselves: calcium. Calcium makes our muscles contract, it is needed to sustain our bones and it is one of the most important substances used when cells want to talk to each other. Mice interestingly have at least two calcium-oriented taste receptors.[70] If salt is a taste, then surely calcium should be one as well. The matter seems further sealed as we humans possess one of these calcium-sensing receptors on our own tongues. Still, it is unclear if it actually helps us taste calcium, at least consciously.

Studies have shown that humans will have an adverse response to it, describing it as quite bitter. Some therefore speculate that the purpose of these taste receptors is not to ascertain a steady influx of calcium, but rather to avoid overconsumption.

Despite the decades it took to implement the concept of umami in Europe, the Japanese company that promoted the taste has recently suggested a new addition to the *Basic Five tastes. Kokumi* refers to a feeling of enrichment.[70] Rather than possessing distinct characteristics on its own, it is described as making the other tastes stand out more. This proposed auxiliary taste was shown in 2010 to interact with certain substances found in food, particularly milt which refers to the seminal fluids of fish, making the tongue's calcium receptor respond more readily to them. Much like the initial response to umami however, this Japanese concept has been slow to be accepted in the Western scientific community. Considering the money involved in the food industry, we can probably expect more time being invested in spreading the use of kokumi. Time will tell if the West will follow suit.

The East-West gustatory divide only becomes clearer as we dive further into this list. Considering how culturally implicit the concept of taste is, this is perfectly expectable. Perhaps we can view the third item on our list as a chance to develop our global cultures a bit. We have talked about the spicy feeling caused by a perceived increase in temperature from eating chilli peppers. In Asia, this is sometimes called *piquance* and is considered a taste on its own.[70] The scientific basis for this would be that there is a receptor for this type of food experience, although scientifically speaking it is really sensing nociception, pain, through its detection of capsaicin. To elaborate a bit more, capsaicin does not increase the temperature itself. Instead, it lowers the pain threshold to heat. Since this affects the entire body, it is no surprise that eating spicy food makes you sweat. Much like catfish have taste receptors all over their bodies, our sensory areas are essentially covered with receptors for capsaicin. So if we consider piquancy to be a taste, then humans too could be considered walking tongues for this particular substance. The opposite sensation, coolness, has also been suggested as an individual taste much for the same reasons as those for spiciness. Coolness might trigger different receptors, such as those detecting menthol, but the general concept of temperature-based taste perception remains the same.

Another sensation related to taste which is familiar to many is the foreboding feeling of having a metallic taste in the mouth. Quite often this refers to the taste of blood. Before the invention of fire and cooking, this might have been sensory feedback on a successful hunt, but today we mostly associate it with something having gone wrong.[70] Nevertheless, many pride themselves in having their steaks bloody, hinting at their caveman origins. Many current foods and pastries are also garnished with some sort of metal, like gold or silver plating. As of yet, we have failed to find any distinct receptor picking up on metallic chemical properties, though metallic tastes can be produced when cerebral parts of the taste systems are damaged.

Considering our appetite for fatty foods, it might not come as a surprise that fat itself has been suggested as a basic taste.[70] We taste sweet things because of the calories most such things contain, so it would make sense if fat, incredibly high in energy, would be put on equal footing. The ruling theory is however that humans cannot taste fat itself but rather enjoy the texture it provides. In this way, *fattiness* is more related to touch than taste, contributing to the amalgamation of senses which generate flavour. The ability to detect fat also seems to differ between individuals. It has been suggested that the reason behind this is reflected in an individual's own level of body fat.[71] Those with higher, unhealthier levels of body fat have been shown to have a reduced perception of sweetness. This may have served to reduce the intake of high-caloric foods, or at least signal satiation.

The last example we will explore is carbon dioxide. Traditionally, the taste of carbonated drinks has been attributed to the prickly feeling they produce on the tongue, which much like pain is relayed through specific channels in the trigeminal nerve.[72] This might be better explained by the multisensory phenomenon of flavour rather than the much more limited scope provided by taste. In 2009, however, it was found that taste-receptors for sour foods express a specific enzyme that detects carbon dioxide, at least in mice.[70] Considering the way humans recognise the bubbles produced in carbonated beverages, it has also been argued that we too might possess this hitherto hidden receptor. Studies have shown that people who take acetazolamide, a drug blocking the enzyme mentioned above, do not experience the carbonated aspect of fizzy drinks.[73] While potentially tangential, this brings up another aspect of taste. A common reason for taking acetazolamide in the first place is to combat altitude sickness when mountain climbing and interestingly, a drop in air pressure and lower oxygen levels may cause altered taste perception; just think about airplane food and how the lower pressure, usually at around 2,000 meters, makes your taste buds go physiologically numb.

These examples of additional *taste candidates* pointedly illustrate how culture and biology intermingle. From having been a subsidy of touch, taste now stands on its own feet in the sensory pantheon, though the ancient discussion of its existence has now evolved to discussions on what we can consider to qualify as taste. The arguments range from the subjective notion that "since we can detect it, it should be considered a taste" to the objective identification of chemical receptors which allow our brains to register the presence of a certain particle. In a way, it may not really matter how we define it. Our brain certainly can detect calcium regardless of whether we are ready to accept its status next to the sensations of salt or sweet. Similarly, in the Japanese kitchen they will continue to discuss how kokumi brings a dish together regardless of what the rest of the world thinks.

Science and culture are intertwined in a never-ending dance. As scientists, it might feel easy to dismiss the subjective views unsupported by quantifiable data but we should remember that our interpretation of that data is based on our own understanding of the world. Of all the senses, taste or its perception is perhaps the most intertwined with cultural nuances. I think it is fair to believe that our sense of taste will continue to be a source of discussion. As it guides us less towards survival and more towards luxury, we can only hope that our culture of high-caloric consumption evolves in a different direction, and quickly.

Chapter 6

Touch

The Catch-'em-all of Senses

Touch I find to be the most complicated sense by far. There are so many different facets to this most basic of senses. When we talk about touch, we mostly mean mechanoreception, as in registering when we feel pressure on a part of our body.[1] Sadly, it is not quite as simple as that. Touch has become somewhat of a catch-all term for the physical sensations that did not make it into any of the other sensory categories. Even taste, as we saw in the last chapter, used to be considered a form of touch before it was given its own platform.

To highlight the multiple facets of what can be considered to come under the umbrella of touch, let's imagine you are walking on the beach and an unexpected wave slithers up your trouser legs. You can feel the cold, salty water as it forces you to move up your next laundry appointment. Thermoreceptors will detect the cold water, while chemoreceptors take note of the salt or any strange chemicals you should avoid. The perception of both temperature and chemoreception come under the realm of touch. Similarly, we possess proprioception, the ability to discern the location of a body part by internally processing where it should be. For instance, when you move your arm, the same motor neurons that fire off the movement send signal copies to other parts of the brain so that all branches of the cerebral "Main Office" know what is going in. Vibration is another type of somatosensory input whose signal is treated differently from that of pressure.

Depending on your perspective, these sensations might all fall into the category of touch, or they may merit separate categorisation. While their physiology differs greatly between them, it makes little sense to differentiate between touching something hot and touching something thermally positive. Nevertheless, this is a topic which is still hotly debated amongst academics. Just like with taste, it might be challenging to change our perception of touch and encompass these other senses under its umbrella when most languages and cultures are built around the Aristotelian Five. For that reason, we will here deal primarily with the most commonly recognised meaning of touch, encompassing mechanoreceptors (sensing pressure), thermoreceptors (sensing temperature), chemoreceptors (sensing chemical properties) and nociceptors (sensing pain). These together make up a concept generally called somatosensation.

The somatosensory system ranges greatly in scope. Humans, sporting a big forebrain and cerebral cortices, have a rather substantial sensory cortex to which all aspects of touch are projected. You do not have to be that evolved to have a sense of touch however. In a way, the sense of touch is necessary for anything to be considered alive, as responding to stimuli is one of the seven characteristics of life. To get some perspective on how the perception of touch developed, let us take several steps back along the evolutionary tree and discuss some pretty inconspicuous life forms that have, in some way or another, been here since the oceans were young.

A Tiny Touch

All animal life stems from one type of organism, the eukaryotes. All cells of your body are eukaryotes, which just means that they contain a nucleus. Pare down the human layers enough and you will end up with this one cornerstone of all biological life. One such lifeform still around today, serving as a reminder of our humble past, is the amoeba. You have probably heard of these minuscule little creatures before. Being small and able to slither around by creating make-shift legs of their entire bodies, they move around unseen to the naked eye.[74] Their sense of touch is what allows them to accomplish this first athletic feat of all biological existence. As the amoeba changes the shape of its outer membrane, reforming its body into a giant foot, the mechanical feedback allows it to reshape its body around surfaces. The hope is of course that this surface corresponds to some sort of food, and the amoeba consequently envelops it. These principles hold true for many of the smaller co-habitants on this planet. For example, orthopteroid insects, like cockroaches or locusts, are equally reliant on touch as perceived through their antennae.[75] The texture of whatever they come into contact with plays a crucial role in deciding whether its owner will approach the object in anticipation of a meal, or run away out of fear for becoming one itself. Another example of why touch plays a key role throughout evolution is its role in synchronising movements. For instance, stick insects compare the position of their antennae to that of their legs, producing an internal movement cycle that allows them to perambulate their thin form forward.[76] Even sponges, lacking any real nerves, have been identified as possessing a sense of touch.[77] If you imagine a sponge, they can be somewhat finger-shaped, reaching for the surface of the ocean floor. In the middle of these fingers, you find a hole which is called an osculum. Water exits through this hole after hopefully having left plenty of nutritious micro-particles for the sponge to absorb in its central cavity. Naturally, this requires some sort of current, which is generated by the sponge as it contracts its body like an aquatic feeding tube. A recent study has shown how small hair cells lining the osculum cause these contractions.[77] The water flows across these thin sensors and creates a feedback loop informing the sponge on how its pumping is going. Much like you can feel the current of a stream or the direction of the wind by sticking your finger out, the sponge uses its hair cells as mechanical sensors for touch, and as we shall see later in this chapter, hair is somewhat of an evolutionary favourite, crucial for many animals' somatosensory perception.

These are all just examples of how simple organisms use touch as a critical survival mechanism. The more interesting stuff comes later in the phylogenetic game, as the eukaryotes pile onto each other forming more evolved vertebrates. Instead of just being a matter of "yes-approach" and "no-avoid" responses, touch allowed us to gather more and more information about our immediate surroundings. We still see this very clearly even in children today, being very tactile in exploring their environments first with their mouths and then with their hands. Research has even shown that pointing, generally involving no physical contact, originated as an attempt at touching something in the far distance.[78]

Mostly About Hair

Touch's all-encompassing nature highlights its evolutionary importance. The various ways through which this sense is utilised hints at a concept we talked about previously when exploring vision: convergent evolution. It would seem all life relies to some extent on touch, and even if they do not share a common ancestor, essentially all animals, and even plants, respond to coming into contact with another object. For animals, evolution seems to have gifted us with hair as the most common method of touch perception. Human hair cells, scarce as they might be in comparison to some of our hairier animal counterparts, all respond to touch. For instance, the whiskers of mice are instrumental in the way they orient their world, and even jellyfish make use of hair cells.[77] Most evidence points to hair having initially been grown to facilitate this touch-oriented perception of the world. Only later did the secondary benefit emerge in the form of heat insulation, which for some animals eventually took over as their main purpose.[77] Most mammals retain this primordial use of hair except for one notable exception. No matter how impressive your moustache is, it cannot be argued that humans use hair as any real method of perceiving our surroundings. Quite likely it was our hands that rendered it superfluous, being far more apt in relaying sensory information. If you think about it, except for us, almost all of our four- legged friends have whiskers in one shape or another. We will go deeper into this later, when we talk about how other animals experience touch.

Despite all the effort evolution has placed into producing a sense of touch, the relative importance of our five senses naturally shifts between species. The eagle might be more reliant on vision than on touch, while a bacteria has nothing but its tactile tendrils to depend on. Humans have moved away from relying on this primordial sense of touch, instead opting for the much safer method of ocular inspection. This happened initially in early primates, leading to an impressive increase in brain volume responsible for visual processing.[79] However, while the visual areas increased with the multiplication of neurons brought on by evolution, the somatosensory regions developed as well. As an indication of where our priorities lie, the cerebral representation of the forepaw increased in particular. We will go into how our body is represented in the sensory cortex later in this chapter, but considering the importance of our hands, I wanted to lead with this example. In some ways, you could argue, this is the birth of what made humans capable of using tools, catapulting us to the top of the food chain, as well as an existence of relative baldness.

The Ways We Can Touch

Before we begin, let us summarise the components of touch as defined by its representation in the somatosensory system: we can feel pressure, temperature, chemical compounds and pain. Combined, these make up what people generally talk about when they refer to our sense of touch, known scientifically as haptic or tactile perception.[80] The former encompasses all active exploration using touch (for instance, hunting for a chair in the dark), while the latter refers to the identification of a touch through passive contact (for example when sitting on said chair and instinctively knowing what it is). Either way, all parts of the human body, particularly the palm of the hand, are specialised in using touch to help create a mental image of an object. The ability to identify objects in this way is called stereognosis and is a rather complex affair. There are a whole bunch of different types of receptor in the skin, which gather different types of data that in turn are relayed to the biological computer that is the brain. Since different aspects of a touch are relayed by different receptors, it feels apt that we go through them briefly, and you can find them represented in Figure 5.

When you push your hand against something, there are two essential components to that touch: the initial contact with the object and the continuous pressure from remaining in that position. The initial light touch is picked up by Meissner's corpuscles. These are the most numerous of the touch receptors. Unsurprisingly, they are primarily located in regions where we have the best sense of touch, such as the palm of our hands, soles of our feet and across the face where they reside close to the skin's surface.[77] Adapted for their need to detect that first contact with an object, the Meissner receptors are highly adaptable. For example, say you are holding onto a heavy bucket with both your hands. As your palms begin to sweat, you risk losing your grip. Thanks to the adaptability of these receptors, that slight change in pressure as the bucket is slipping between your sweaty palms allows your brain to register what is about to happen. Subconsciously you tighten your grip, saving the precious contents of your bucket. The Meissner receptors are four times more sensitive than the second most numerous kind, the Merkel corpuscles. These receptors respond to continuous pressure, and predominantly residing in the tips of our fingers, they give us the ability to discern shapes like angles and curves of an object.[19]

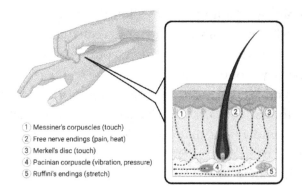

1. Messiner's corpuscles (touch)
2. Free nerve endings (pain, heat)
3. Merkel's disc (touch)
4. Pacinian corpuscle (vibration, pressure)
5. Ruffini's endings (stretch)

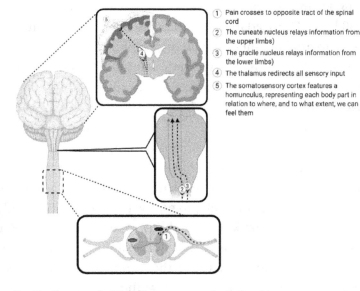

1. Pain crosses to opposite tract of the spinal cord
2. The cuneate nucleus relays information from the upper limbs)
3. The gracile nucleus relays information from the lower limbs)
4. The thalamus redirects all sensory input
5. The somatosensory cortex features a homunculus, representing each body part in relation to where, and to what extent, we can feel them

Figure 5. Feeling touch. This figure on touch deals with somatosensation, which one may also call external touch. The top illustration shows the structures involved in this process, with the specific domain of each receptor outlined in parenthesis. The arrows indicate the electrical impulses created for our central nervous system to interpret. This latter process is outline in the bottom schematic, from the crossing of the central line to them reaching the somatosensory cortex which allows us to perceive the signals as touch. Created using Biorender.

As such, they are slowly adapting so as to allow the brain some extra time to work out what its bodily appendages are actually touching. They are less exposed than their Meissner counterpart and hide away in the deeper layer of our skin.

The Meissner and Merkel receptors consequently give us the all-important ability to first detect that we have come into contact with something, and secondly what that object feels like. There are two more basic receptor types that add an extra finesse to this basic perception of touch. Since the Meissner corpuscles detect light pressure, they can also detect slow vibrations, which in a way are just a light pressure played on repeat. Proper vibrations are however caught by Pacini's corpuscles which are sensitive to frequencies of up to 350 Hz.[19] These adapt faster than the Meissner receptors since they need to update their signal very rapidly. The onion-shaped Pacini corpuscles also have a very low activation threshold, sensing vibrations as small as 10 nanometres. The downside is that their receptive fields need to be quite large and we therefore experience vibrations as rather diffuse and hard to locate. Arguably, the purpose of this aspect of touch therefore seems to be less about locating the exact location of a vibration but rather identifying that there is one, which for an early human quite possibly could have been brought on by a stampeding elephant, for instance. Today, we have Pacini's receptors to thank for our ability to do tasks requiring fine motor skills. Writing with a pencil, for example, produces small vibrations that are transduced into the hand. Being able to sense this, the brain can adapt to safeguard that the vibrations do not grow too large, ensuring that the written lines stay thin and in the right position.

The last type of the basic touch receptors does not really have much to do with the somatosensory system. Rather, it allows us to know the location of our body parts through what is called proprioception. Seeing as these distinct receptors neighbour their somatosensory siblings, I thought it suitable to at least mention these structures known as Ruffini's corpuscles. These cylindrical receptors are not only found in the skin but also in tendons and ligaments.[19] Laid out in the direction of the skin, Ruffini's corpuscles react to stretching movements. As you move your hand to turn a page, these stretch receptors activate in order to signal to the brain that a body part has moved. The brain in its eternal wisdom calculates the new position of the limb so that we do not end up forgetting where we put that left hand of ours. These are also slow to adapt, which is important since they need to continue signalling for our brain to know where the body part is. Leave a leg rested in the same position for some time, however, and you will have no conscious notion of its angle. This is because Ruffini's corpuscles have adapted to this new position and will not start firing again until a new movement takes place. Had this adaptation process not happened, we would probably have had a pretty hard time ever relaxing as no position would feel natural.

Creating a Signal from Touch

These four specialised types of receptor are all mechanoreceptors. Having picked up respective aspects of touch through their receptive capsules, they soon pass the signal on through their neural axons in the form of electric impulses. These axons, the long winding part of any neuron, are relatively thick, about 6–12 micrometres in diameter.[81] Physiologically this is of huge importance as a greater thickness means higher speeds, and the electric signal in this case travels at 33–75 meters/second. As a point of comparison, this would put it on par with the speed of your average train, but only make it half as fast as the Shanghai Maglev Train, or even a third of a passenger airplane; I hope this can be of assistance should you ever find yourself in a discussion regarding "the speed of thought". In such a case, you can point out that there is much room for improvement. Comparing this relay network to artificial sensors, consider that electrical currents travelling through a human-made copper wire travel at 95% the speed of light, vastly outdoing our human reaction time. The body does try though, and in addition to the thickness of the axons, the way they are covered in a substance called myelin helps the impulses move along.[81] Consisting of a fatty substance, myelin is organised in sheaths that act as insulators for the neuron. Between the many segments are tiny little pieces of exposed neural tissue called nodes of Ranvier. The touch-signal hops between these nodes, allowing it to jump large sections of the neural highway so that it may reach the brain faster.

We earlier said that pain and temperature work differently from their mechanoreceptor cousins. While pressure receptors live in an enclosed capsule, pain and changes in temperature are picked up by ragged, free nerve endings that end near the hair follicles.[19] These nerves lack the covering myelin clothing, and are about 12 times thinner than the named touch-corpuscles. These two factors, the lack of myelin and their relative thinness, make it so that pain and temperature are relayed much, much slower than the other touch signals. Travelling at only up to 2 meters per second, the pain impulse would lose a 100 meter race against Usain Bolt by about 40 seconds.[81] This means that if you were slapped by a short distance runner, he or she would be pretty far away before you could even register the pain. This has some interesting implications. Consider a young child having scraped a knee. With tears in their eyes, the child whimpers slightly as a parent gently blows on the wound. This might be a familiar scene for many, and the reason why parents do this goes further than placebo. Due to the much higher velocity of the Meissner receptors compared to the pain neurons, a signal of the concentrated wind will reach the brain before that of the wound. As a result, the light touch of the parent's gentle blowing will be felt before the pain, giving it priority in certain brain areas. This hotwiring of the somatosensory system has given us a cheap analgesic while also highlighting the complex neural network that makes up our sense of touch.

In order for an electric impulse to be produced, enough pressure of the right kind needs to be applied to a receptor. We have already mentioned that the reason parts of our body, like our hands or face, are more sensitive than others is because of the high concentration of touch receptors found there. Our body parts also differ in that they have different receptive fields.[1] You can think of these as a very sensitive fishing net, which the fisherman then collects after enough fish have been caught. In a similar way, different receptors have different sized nets, requiring different types of contact before they fire an impulse. This net is 2 millimetres (mm) on the fingertips, about 40 mm on the back and roughly 47 mm on the calves — should you be in the market for a tattoo and worried about pain, then this might give you an idea of where to start.[19] As we said before, when the sensory catch has been reeled in, it produces an action potential, the electric impulse, which is fired through the axon.

Before we follow this impulse into the spinal cord, we need to talk a little bit about proprioception, which you can think of as an internal sense of touch. We stated that one of the four named receptors, Ruffini's corpuscles, are important for the conceptualisation of how the brain organises its own position by measuring how much a certain limb has stretched. There are three more general types of mechanoreceptors that aid in this process. These all share a common love for being activated and consequently have a very low activation threshold, meaning that even the smallest of muscle contractions or joint adaptations will be registered in the brain. Muscle spindles are a type of stretch receptor that, unlike Ruffini's corpuscles, reside in muscles, especially those important for postural control and balance.[1] They also make sure that all muscles responsible for fine motor control behave as they should, like the hand, the tongue or the eyes. They are not especially small, but fuse well with the surrounding muscle tissues. At about 4 millimetres in length and 1 millimetre in width, they fuse with the internal structure of the muscles in sharp-tipped cylindrical shapes. These come in two flavours, type Ia (primary) and type II (secondary).[19] Primary receptors are rapid adapters, much like their Ruffini cousins, and signal when a movement is initiated and when it has stopped. The secondary type signal constant position of a muscle group and therefore need to adapt very slowly. So in short we, through these receptors, are able to write a short story on how our muscles are behaving. The journey of a muscle is told by the primary muscle spindles, writing down every turn and tumble a muscle experiences.

Nestling in with the Ruffini receptors in the joints are the second of the designated proprioception transducers. The Golgi tendon organ connects the end points of a muscle tendon.[19] These are thick and dressed in a coating of myelin, making them quite fast. When the muscle contracts, the Golgi receptor fires its action potential as the tendon shortens. It works a bit like a luggage scale: hold up the muscle in a hook and the scale will give you a reading of how strong the pull is. As they adapt slowly, they continuously inform the brain on how well the muscle is doing during its contraction. The last type of proprioception receptor is the joint receptor.[19] It acts in conjunction with the Ruffini receptors in the elbows, knees, and other moveable examples where bone connects to bone, and provides information of a joint's angle.

These three receptor systems, or homologues of them, exist both among vertebrates and many simpler lifeforms. In summary, the tools embedded in our skin, spying on the surroundings on behalf of the brain, are the Meissner, Merkel and Pacini corpuscles, dealing with light touch, pressure and vibration, respectively. The Ruffini receptors are also found in the skin and respond to the stretching produced by contact with an object, relaying information on how the skin has moved. This can be considered a form of internal touch, or proprioception, which is also consolidated by the work of muscle spindles, Golgi tendon organs, and the joint receptors.

Many Ways to the Brain

At long last, we are finally getting ready to send the signal onwards to the brain. The sensory signals from the different types of touch receptor will travel along the axons into the spinal cord, where they enter from the back. Since touch is comprised of so many different types of signal, it might not come as a surprise that they are divided here as well. While touch signals will go straight to the brain, pain and temperature need to be processed in what is called Rexed's laminae.[1] This is important in the example of the child with the grazed knee being comforted that we talked about before. The pain axons go into Rexed's first layer, relayed by two different subtypes of neuron that respectively convey dull pain which is hard to localise, and sharp pain that can be pinpointed. Layers III and IV integrate many of the other pain and temperature signals before they are sent higher. Going back to the example of the child, their pain is conveyed through both dull and sharp pain axons. A gentle blowing hopefully only activates the more benign touch receptors. Since these interact in the Rexed's laminae, the integration is such that the blow-signal downregulates the pain through inhibiting its signal.[1] Known as the "Gate Control theory", this interaction is hugely important for treatments and research on pain. Some aspects of painless touch, from muscles and joints, are also integrated in Rexed's laminae. The reason for this is to quickly provide an idea of how the body is moving that can be immediately sent to the brain's fact-checker, the cerebellum. The cerebellum immediately responds by altering the muscle contractions to make sure that the brain's command is being followed.

Signals in the shape of light touch and vibrations do not mix with their painful cousins. While they also enter the spinal cord through the back, these axons move straight into the elevator that will bring them up into the brain: the funiculus posterior[1]. Most signals travel straight up on the same side of the body on which they entered the spinal cord. The funicular elevator takes the signals as far as the medulla oblongata. We are now in the lowest parts of the brainstem. Here, the signals split up in a wonderful example of how the physical wiring of our nerves helps the brain categorise its many areas of responsibility. The ones which stood in the middle of the elevator, the medial axons, represent the lower limbs and go through the nucleus gracilis. The lateral ones, having stood to the side, instead carry information from the upper body and go to the nucleus cuneatus. From here, the signal crosses the body's midline. Information which has come from the left arm, for instance, goes into the right brain hemisphere and vice versa. Before they move on into their respective target areas, the signals will, as usual, be filtered through the thalamus. From here, a third set of neurons carry the information to the primary (S1) and secondary (S2) sensory cortices. Perceptually, this is where your subjective sensation of touch takes place.

The primary sensory cortex ends up accepting the majority of the somatosensory signals, as illustrated in Figure 5. It is located in the parietal lobe, almost at the centre of the tip of the head[1]. Keeping in line with subdividing aspects of touch, S1 is compartmentalised further into four regions collectively known as Brodmann areas. All of these sections contain a complete representation of the human body through a well-defined local neural network. These areas are however highly prejudiced; they generally over-represent the hands, lips and face, while only providing a small mention of the lower extremities, except for the feet which seem to have climbed above their stature. This model of representation is called a Homunculus. The word itself describes its purpose rather well. Meaning "small human", the Homunuclus' representation reflects the way we perceive our various body parts. Rather, we perceive them as we do because of the shape of the homunculus. At any given time, you will have better sensory control over your fingertips than the small of your back. You can feel the oncoming blister on your lip much better than a day-old bruise on your thigh, unless you have kept prodding it. Simply put, the brain prioritises certain sensory body parts over others. We need excellent sensory feedback on our fingers if we are to use the tools that have catapulted us to the top of the food-chain, and knowing the surface on which we place our feet is obviously advantageous.

After having integrated the information, the S1 cortex is ready to let the signal move on to the secondary somatosensory cortex (S2). Here, the signal enters relatively indiscriminately compared to in S1. Still, the skin's perceptive fields continue to be presented in this region, and the already filtered touch-signal is getting ready to be finalised into something that can be felt and acknowledged by our own consciousness. In line with the brain's holistic agenda, this integrated signal is also projected to other brain areas so that an appropriate response can be constructed. For instance, the limbic system, responsible for our emotional responses to external stimuli, projects the signal into the amygdala and hippocampus in order to make sense of it. The amygdala might reveal that the touch is from an escaped deadly caterpillar, producing a fear response and forcing you to shake it off. Meanwhile, the hippocampus acts as a very astute bibliographer, cross-referencing this most interesting signal to previous experiences in its library, aiming to create an effective memory of touch.

Touching on the Clinical Perspective

With so many steps between the external tactile stimulus and its final cerebral endpoint, it might seem incredulous that we are able to feel anything at all before any tactile threat has gotten the chance to eat us. Thankfully, there are some shortcuts. S2 also sends an immediate signal to the motor cortex, where all our movements are constructed.[1] This is how somatosensory information and motor actions are coordinated. If the aforementioned caterpillar were to throw itself at you from a tree branch, you would reflexively dodge away from it thanks to this shorter circuit. Unfortunately however, the ability of the sensory system to learn is also what makes it end up playing tricks on us. Since it works together with the motor system, creates memories with aid of the hippocampus and represents specific areas, it is very sensitive to changes that arise in this system and when trying to account for something new the brain area adapts

after a while. Phantom pain is one example of how this could manifest. As the brain struggles to make sense of the loss of a limb, its
sensory representation in the brain is consequently taken over by
the adjacent area, meaning that the hand now answers for what happens to the non-existent finger. A touch on the palm of the hand
can instead be interpreted as a prick on the finger. So, even though
the finger haunts its previous owner with its absence, it makes itself
known through this rewriting or rewiring of the somatosensory
cortex.

The long neural network from skin to actual perception in the
brain also leaves it vulnerable for some interference on the way.
Vibrations, pressure, light touch and proprioception make their way
to the brain using the pathway we have just gone through. Pain and
temperature take a different path up the spinal cord.[1] While these
signals enter on the same side as their benevolent brethren, the
dorsal horn and Rexed's laminae, they immediately pass on to the
other side of the body's midline. Here, they hike up the spinothalamic tract which is located straight across the vertebrae, facing your
stomach. This creates a short of X-shape if you take both sides of the
body into consideration, where light touch stays on the same side
and towards the back, and pain travels across towards the stomach.
As the name of the tract implies, it takes spinal pain and temperature signals to the thalamus, from which it is indexed and sent
onward in much the same way as the other forms of touch. Some
interesting phenomena may occur due to this crossing of neurons.
Spinal cord injuries are sadly not uncommon, causing a range of
different symptoms depending on the location of the lesion, herniation, or other type of trauma that has altered the neural highway.
Let us assume someone has a spinal disc herniation just a couple of
centimetres underneath the navel, damaging the tissue on the right
side of the vertebrae. As some axons travel on the same side of the
spinal cord as the limbs they innervate, this would mean that the
right leg will cease to provide the brain with information regarding
vibration, pressure, and light touch. The receptors in the leg still

work and will continue to pick up any sensory information. However, since the brain never receives the signal, it is as if it never happened. The only type of somatosensory information that will reach its intended target is pain and temperature, which having crossed over the spinal cord are therefore not affected by the damage. Consequently, while the right leg has lost its sense of pressure, it still feels pain. The left leg has instead become impervious to pain while having no problem relaying light touch information. Testing touch and pain in a patient is therefore a very efficient way of discerning the level of a spinal cord injury. Each disc also receives information from a specific region of skin called a dermatome, in line with the segmented wiring of the neural network. From a muscular perspective, we see essentially the same thing, although now the region is called a myotome. Considering that investigating muscle function and skin perception reveals a lot of hidden information about the spine, this is why doctors perform seemingly unrelated poking tests whenever they want to rule out neurological problems.

As we stated at the beginning of this chapter, the sense of touch is a complicated one. Apart from the example above, there are several means in which our sense perception can be used to diagnose different pathologies. Tactile hallucinations, the perception of a touch that has not actually taken place, are not uncommon and may be due to tricks played on us by our peripheral nervous system as well as issues with the brain's internal processing.[1] There is also something called referred pain, when damage to one part of the body is perceived as having originated in a different limb. It is for example quite common to experience pain originating in the heart as a discomfort in the left arm, neck or jaw. Anatomically, we can trace this back to how the nerves come together before entering the spinal cord. Damage to a nerve after this conjoining means that the brain will perceive the damage to come from all sensory innervation areas that the jumbled up nerve bundle receives information from. Facial somatosensory innervation poses another deviation from the pathway we have just gone through. Since cranial nerves enter the skull after the spinal cord has ended, they have to follow their own path, which for somatosensation happens through the fifth cranial nerve, the trigeminal. We have talked about this in a previous chapter, since this pathway is involved in how aberrations in light, smell and taste perception can cause physical pain. While there are animals that are blind, deaf or absorb stuff into themselves without even acknowledging the concept of taste, the sense of touch serves as a universal reminder of our common ancestry, a phylogenetic agreement on what constitutes the bare necessities of life. From the deprived amoeba to the great apes, touch remains our most basic denominator.

The world of touch is clearly a widespread one, ranging from the highly specialised neural network of mammals to the simpler mechanical actions of spongy hair cells and floral leaves. As the oldest sense, it has guided all aspects of life on this planet since the first strands of RNA bound together in the stormy oceans billions of years ago. It has been argued to be the sole human sense, the foundation on which the other four rest. Even Socrates himself viewed taste as a form of touch and I personally think that a point can be made for our sense of hearing being considered a variation of the somatosensory perception of vibrations. You could of course argue the other way around; the different aspects of human touch are too different to be considered part of the same sense, although this would make describing what a touch means much more awkward — just imagine, having to describe the touch of a loved one in terms of vibrations and thermal fluctuations would make it lose some of its intrinsic sincerity.

Chapter 7

The Hidden Senses

Hidden in Plain Sight

By bringing up our set of human abilities that could be considered senses, I set out to achieve two things. The first one was intentional: to present some alternative views on how we can think of our own capacity for perception. The second was not making enemies of any scientist whose field of expertise I have left out in the main sensory chapters. We certainly have more senses than the mere five we have dealt with thus far, but worry not because we are about to widen our scope. Let me however start this segment by saying that I have no personal preference on what a sixth sense should look like. As we saw in the introductory chapter, our notion of the senses has developed more as a cultural phenomenon rather than merely through pure scientific endeavour. Plato referred to desire, pleasure and despair as internal senses, and we could certainly spend some time in this chapter discussing the physiology of love, but this has been covered quite well by many other writers so this chapter will instead try to give a platform for the less privileged biological functions. As much as they all deserve a platform on their own, this chapter seeks only to introduce some of them briefly, and to some extent, champion their cause for being considered a proper sense.

Our sensory systems reflect key aspects of the way the earth is built up. We have hearing because there is an atmosphere in which air can vibrate. We have touch because there are objects more solid than the aforementioned air that we can to orient around. Often, these aspects are so fundamental we hardly notice them. The way vision produces an image in our brain is such a quintessential aspect of human life that we rarely give it enough credit, yet it is based on something as fantastic as absorbing the reflections of a fusion reactor suspended millions of kilometres above our heads. Imagining a society where a different sense would have been the main source of inspiration is an effort in futility. What if buildings were designed to reflect soundwaves rather than for visual and tactile practicality? Such a question is of course pointless, because in such a case we would have had evolved to prioritise that sense much in the same way our brain covets vision. Either way, the reason we think of vision, touch, and the other senses as basic necessities for human civilisation is because they were there from the very start of our existence.

We will start off this chapter with another such primordial physiological concept. It was there before vision, maybe even before touch. As earth was forming, not even sunlight could penetrate its thick gaseous atmospheres. Still, the rumbling core of the planet presented all life-to-be with an obstacle so basic we take it for granted: gravity.

A Balancing Act

In practical terms, the existence of gravity means there must be an Up and a Down. As a consequence, all organisms inhabiting this planet must adhere to this directional bias. This is where our first candidate for a sixth sense enters: balance, or equilibrioception. This involves all types of movements that somehow involve gravity: movements, acceleration, the position in which we are situated and the direction of any of the aforementioned activities. These critical concepts are certainly worthy of mentioning when talking about important information we gather about our surroundings. Still, what is the physiological background for this ability? We have eyes, ears, skin, a nose and mouth, but where do we look to find an organ that physiologically backs up this sensory claim? We actually do not have to leave the head, but we do have to go a little bit deeper than what meets the eye, skin, and ears.

If you were to follow your ear canal, you would eventually find the cochlea, the snail-shaped structure which interprets soundwaves, as discussed in the Hearing chapter. This is roughly located a centimetre or so beneath the outer corner of the eye. Despite its burrowed location and the cochlea being an elongation of the ear's function, this is not what we colloquially refer to as "the inner ear". Jutting out from the spiralling shell are three rings, reaching in three different directions, as illustrate in Figure 6. This is the inner ear, also known as the vestibular organ. Those three rings, called the semi-circular canals, are responsible for detecting rotations of the head in the three dimensions our bodily forms are limited to inhabiting.[1] Jutting out at right angles from one another, they are justifiably named the posterior, anterior, and horizontal canals, and are able to sense all possible directions your head might end up in. At the base of these three canals, we find two bony pouches containing tiny crystals. Attached to tiny hairs (by now, you will have learned the importance of hair cells), these crystals tug on the tendrils as our head changes position. These pouches, the saccule and utricle, are called the otoliths. The saccule, with hairs pointing horizontally, is sensitive to positional changes in the vertical direction. Inversely, the utricle's hair cells jut out vertically and are sensitive to horizontal movements. Altogether, the inner ear presents a complex mix of motion detectors. These two features, the semi-circular canals and the otoliths, are part of the physiological system solely dedicated for our perception of balance, which you can argue allows us a perception of gravity. As we have done with all the other senses, all of them similarly complex in their own way, let us start where the nerve begins.

It all begins with a movement of the head. The fluid in the semi-circular canals, the same type of endolymph that we find in the

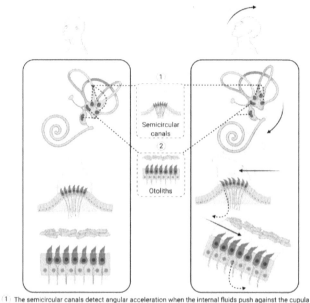

1) The semicircular canals detect angular acceleration when the internal fluids push against the cupula

2) The otoliths of the utricle and saccule signal head displacements when the otoconia crystals drag on hair cells

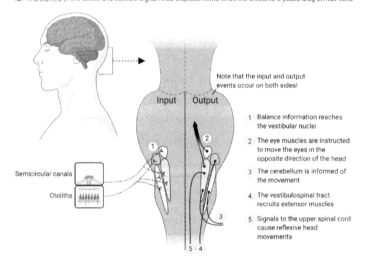

Figure 6.   A balancing act. Our sense of balance is extremely important, as it allows us to remain upright and retain steady vision. The top image illustrates how the peripheral vestibular organs detect movements and changes in posture. This information is converted to electrical impulses through mechanical forces acting on hair cells, and the signals are conveyed to the bottom illustration. Here you can see how different parts of the vestibular nucleus contribute to us maintaining control of our bodies as we move around the world. Created using Biorender.

cochlea, stays stationary while the tight walls move around it.[1] To visualise this, think of how you can turn a cup of tea around: if you do it quickly, the tea will stay in place, so if you want to catch a loose leaf you will have to move rather slowly. The vestibular organ is not immune to the same Newtonian laws of motion that apply to the cup. At the end, or beginning, of each of the three tunnels, we find a thin membrane (cupula) blocking the way, located in a bony ampulla. As the fluid moves, the cupula flutters in the endolymphatic stream, pulling with it many hair cells attached to the bottom of the structure. In reality, we can arguably claim that the endolymph does not push against our cupula as much as we actively push the cupula into the endolymph. Much like the way sounds are interpreted, the tugging of the hair cells on their anchored cells opens or closes ion channels based on their direction. In a purely mechanical fashion, the movement has been translated into an electric impulse. This particular impulse is picked up by the vestibular nerve, but it does not travel alone. A movement of the head means a positional change, and while the motion itself was calculated by the semi-circular canals, it is up to the otoliths, the utricle and saccule, to produce an estimation of the new angle.[1] The little crystals we mentioned before play a key role in this. These otoconia are embedded in a gelatinous layer, resting on top of hair bundles. When there is a pull on the otolith hair cells, we once again get an electrical impulse being produced. Based on the angle of the pull, the brain can determine the new position of the head. Much like we can become accustomed to a certain frequency of sound, a static position of the vestibular hair cells will after some time lead to an acclimatisation so that for instance, if we travel in a bus at an even acceleration, we will stop feeling the movement after a couple of seconds.

One of the most common causes of vertigo arises when the two systems come into contact with each other. Crystals from the utricle can be shaken off from their hair cells and fall into the vestibular vestibule, making their way into a canal. As the head moves, the additional weight in the endolymph exacerbates the forces upon the cupula, drastically enhancing the signal in the finely tuned vestibular apparatus, causing benign paroxysmal positional vertigo (BPPV).[82] If you have ever experienced this condition, or any other form of dizziness for that matter, you will probably agree that the vestibular apparatus deserves as much recognition as the nose or ears in the role it plays for how we function in the world and how we perceive it. It coordinates much more than we think, always keeping us aligned with the planet itself. In fact, without this organ you would be functionally blind, as we will discuss below.

First of all, we have two inner ears that work in tandem according to a push-pull system. What this means is that as you turn your head to the right, the right vestibular apparatus will send an excitatory signal to the brain, while the opposing left system will instead be inhibited.[1] Depending on the relative strength of the signal, the brain can decide which direction the head has moved in. Each apparatus relays its signal through the vestibulocochlear nerve, the eight cranial nerve, which projects to the vestibular nuclei in the brainstem. Here, a series of coordinated muscle contractions are triggered in response as can be seen in Figure 6. Our ability to stay upright requires immediate and subconscious motor actions just like these, as we cannot afford ourselves the luxury of realising we are falling before coordinating our legs to adopt a new position to prevent that fall since our conscious perception of things is far too slow for that. It is signals issued from the vestibular nuclei through the vestibulospinal tract that keep us standing. A positional change will cause the vestibular apparatus to signal to the legs through the lateral part of this tract, forcing the knee to extend on one side, with a coordinated flexion of the other. Medial nerves harmonise the upper body in relation to the new position, which crucially involves the neck muscles so that we are able to maintain our heads fixed in space even if our body moves beneath it.

To illustrate another important aspect of the vestibular system, I have encountered a patient in clinic who complained that she had tremendous difficulty enjoying her evening television shows. She claimed the picture was jumping up and down and the initial suspicion was that there was something wrong with the eye muscles, making them twitch unnecessarily, therefore making the image move on the fovea. In reality, the problem was almost the complete opposite, with the eye muscles being perfectly healthy. The issue lay in her vestibular system. The vestibular system normally sends a signal directly to the centres responsible for eye movements.[83] As we move our head to the left, our eyes move the same distance to the right so that we can focus on a visual target. This patient had had an inner ear infection which left her without this vestibular eye-head coordination. The jumping television set was in fact due to the patient's own heartbeats! Sitting down, the head was relaxed enough to be gently pushed with each burst in blood pressure caused by the contractions of the heart. This is what I was referring to earlier when I mentioned we would be functionally blind without the vestibular organ. Without the ability to coordinate our eyes, it would be near impossible to sustain visual fixation on a target. Although some learn to live with it, it highlights how important it is that we are able to perceive our own equilibrium. Why then, considering the impact this perception of gravity has on our lives, do we not consider balance to be a proper sense? Most likely it is such a basic concept that we have just taken it for granted. It does not really fit into any cultural expression either — you could potentially proclaim that your sweetheart makes you dizzy with love, but the semi-circular canals do not lend themselves to poetry the same way the eyes might.

Heed the Warnings

Despite its role in our lives, we rarely view pain, or nociception, as a sense on its own. Together with temperature, or thermoception, it finds itself subjugated to the whims of touch, most likely because of the way both pain and temperature receptors are all jumbled up in the skin. As we have already seen, the physiological mechanisms for these two sensations differ greatly from somatosensation. Let us consider the premise of allowing sensory status to these two modes of perception.

It might come as a surprise to hear that pain is a rather new concept, at least from a distinct neurophysiological point of view. Obviously pain itself has existed for quite some time, but it was long believed to be caused by our otherwise benign touch receptors becoming strained beyond capacity. Only in the mid-20th century was their malevolent twin discovered.[84] As we have already seen, the supporting neural network turned out to be quite different from the somatosensory one, extending from the receptor type all the way to where it is conceptualised in the anterior cingulate gyrus of the brain. Such a long winding way naturally means there are several points at which something may go wrong and cause pathological pain. To understand how this may arise, we need to classify pain depending on the way it is perceived, and scientists generally agree that there are five distinct subtypes of this internal alarm.[85]

The usual type of pain, for example when we scrape a knee or get a splinter, is called nociceptive and is caused by local inflammation in the tissue. This signal is not necessarily even picked up by proper receptors, but instead free nerve endings reacting to inflammatory particles.[1] There are three types of nociceptors: mechano-nociceptors reacting to mechanical stimulation like blunt trauma, chemo-nociceptors that go haywire for instance when you get chemicals such as soap in your eyes, and thermo-nociceptors that cause acute pain at temperatures below 15 or over 43°C.[86] Apart from these nociceptive pain types, there are four others. Neurogenic pain emerges when there is damage to the neural system itself, such as when a herniated disc presses on the neurons in the spinal column.[85] Idiopathic pain is a sort of catch-all term and refers to any pain that lacks a clear cause or biological aetiology, with fibromyalgia being an example. Naturally, this does not mean the pain is any less real, just that we are yet to discover its cause. The fourth type of pain is the mixed type, which is rather self-explanatory. Pain is by definition subjective, and what is painful to one might not be painful to another. Pain can, of course, be one of the mind, and considering how we are all but manifestations of our brains, the fifth pain type, psychological pain, is as real as any other, and can be severe in very common psychiatric conditions such as depression or anxiety. Much like nociceptive pain is a sign to step away from a painful stimulus or do something to rectify the damage, psychological pain can remind us of the impact of stressors on the brain.

Despite its evolutionary use, most of us actively try to avoid pain. While it may seem advantageous to remove this sense from our current lives, doing so would be exceptionally dangerous. For many people, this lack of pain reception is a reality, but rather than presenting an evolutionary advantage, 20% of them die by the age of three.[87] *Congenital insensitivity to pain with anhidrosis*, or *CIPA* for short, is a genetic mutation most notably seen in communities adopting consanguinous marriages. A mutation of a specific gene causes the pain receptors to not function properly.[87] As a result, the brain receives no information on either pain or temperature. Deprived of these essential tools, the brain is incapable of regulating the internal thermostat. This is why so many die young as incapable of sweating, they succumb to hyperthermia. For those who survive, the only form of treatment is to prevent infections and other health hazards by regularly inspecting the body for injuries. A broken bone might never heal since its owner continues to put weight on it, and a small stone in the shoe might destroy the foot due to skin breakdown. To sum it up, pain exists for a reason.

Just as pain is not really a specialised form of touch, temperature is not a subgroup of pain, although in the Smell chapter, we noted how capsaicin, the peptide that makes peppers feel hot, does activate the pain receptors in the mouth to create the sensation of spiciness. Although hot and cold can feel painful, our bodies usually move within much finer confines on the thermometer. Despite this, I find that both pain and temperature share some common ground in how they could be perceived as individual senses: sensory information is gathered primarily through the skin, signalled to the brain through the same spinothalamic tract in the spinal cord, and information is gathered on our surroundings that are often, and unfairly, attributed to touch.[1] Being a very modern creation from an evolutionary perspective, the human body is fittingly equipped with central heating. To be fair, our hypothalamus might not actually do so much heating and is more akin to a responsible landlord, gathering information on the body through its heat receptors, which dip their toes into our blood stream to make sure it is a pleasant temperature.[1] Much like an apartment building, the human body is equipped with state-of-the-art piping that is used for both heating and cooling. It achieves this primarily by altering the diameter of our blood vessels. If we are too warm, the blood vessels in the skin will dilate and increase their surface area so that heat can escape through the tubular walls. A constriction will instead allow less space for the heat to escape, increasing the body temperature.

Such small shifts in area might not seem like much in terms of regulating the temperature of an adult human. Usually however, it is surprisingly efficient. We have such an extensive network of blood vessels that if they were all laid out in a straight line, they would go on for about 100,000 kilometres.[88] The range within which the hypothalamus keeps our internal temperature is significantly narrow, about 1°C around our standard temperature of 37°C.[89] Apart from the purely physiological intervention, our behaviour is also heavily impacted by the hypothalamus. If you find yourself beneath a scorching sun on a hot summer day, you will feel thirst not only because of dehydration but also to help lower your body's temperature. Inversely, the only thing you might be thirsty for in a snowstorm is a cup of hot chocolate. You do not have to learn to put on warm clothes when it is cold, this is instinctively something you know to do thanks to your hypothalamus looking out for you.

However, no system is infallible and our internal thermometer is no different. We are all familiar with stories of ill-fated wanderers getting lost in the wilderness, eventually succumbing to the cold.

What might strike you as strange is that in up to half of those cases, those found had dressed down completely.[90] This phenomenon is known as paradoxical undressing, and acts as a reminder of the fallibility of our own senses, as well as the power that our body holds over our minds.[91] Physiologically, there are two main theories on why this occurs. One argues that the hypothalamus malfunctions during severe hypothermia, defined as having a body temperature below 28°C. The other points to the exhaustion of all the muscles responsible for contracting our blood vessels. If that is the case, a sudden dilatation of the blood vessels would lead to an intense surge in blood flow, tricking the body into believing the temperature to be deceivingly high. Either way, people find themselves adapting to this false perception of temperature despite the colder reality they find themselves in.

You might find yourself saying "well, now that I am aware of it, I wouldn't be silly enough to undress in the middle of a snow storm". So, let us take an example that probably hits a bit closer to home. If you have ever taken a shower, you probably had to adjust the temperature of the water. It might have felt almost scolding at first, but after some time you eventually habituated the thermoreceptors in your skin. Slowly, the signal they were sending to the brain became weaker, indicating that the water might not be so dangerous after all, and that the temperature you first considered painful was in fact not so. You might know this from the start, but your hypothalamus seems to have a pretty bad memory. Some are however said to have the ability to suppress this biological mechanism in a truly impressive case of mind-over-matter, allowing them to walk on glowing embers.

All in all, there is a lot of sense in considering pain and temperature to be methods of perception to be senses in their own right. Maybe touch should be considered a subtype of pain instead, considering how it is given a lower internal priority? I find that both these sensory inputs share some common ground in how they could be perceived as individual senses: tactile information is gathered primarily through the skin, signalled to the brain through the same spinothalamic tract in the spinal cord, and information is gathered on our surroundings that are often, and unfairly, attributed to touch.[1]

Use Your Common Senses

So far, we have covered three systems that are frequently brought up as candidates for additional senses by modern scientists. They are all distinctly physiological, with specific receptors and internal processing mechanisms that could, and have, filled entire encyclopaedias. Still, we rarely talk about a whole group of senses that were seemingly accepted in the ancient and medieval eras. The internal wits were once thematically linked to what would become the external senses, yet we rarely think of them in that way today. Still, if a sense is by definition a system used to gather information on the world, then surely our brain's ability to detect its contents should be celebrated as one. In a way, this ancient view has survived in modern English, even though we might not reflect much on it. The common sense may be the ultimate surveyor of the world around us. The term has almost become somewhat of a platitude, a term only used when berating someone for their foolishness when they lack it. Still, the name remains. Any attempt at elucidating the senses without addressing a concept carrying the phenomenon in its very name seems simply irresponsible, and an author of such a work would surely lack the very thing we are currently referencing. In an effort to appear sensible, let us therefore dig a little deeper into the concept of human intelligence.

Unsurprisingly, we find ourselves once again with the old Greeks. Aristotle himself has been credited with the concept of a common sense as he discussed it in his work De Anima, or On the Soul.[9] The idea was to formulate a more holistic interpretation of what he called the specific senses: sight, hearing, smell, and touch, including the subspecialty of taste. To create a smoothly categorised view of the world, there must be one common integrator, something that creates a joint output of all physical perceptions. Aristotle explained that all animals, not only humans, were gifted with this ability to make sense out of our surroundings. After all, a deer can identify poisonous berries from their look and smell based on previous experiences. A dog can learn to avoid certain individuals, having correlated their scent and appearance with a negative and painful previous encounter. From this perspective, our sensory perceptions are clearly all put together into a general, common sense.

What separates us from other animals, according to Aristotle, is instead the level at which we utilise this sense. Just as eagles have superior vision and dogs unsurpassable noses, Aristotle argues that humans possess a specialised form of the common sense, intellect. Like his contemporaries, Aristotle was fascinated with the internal workings of the mind. Our individual senses could never be much more than a rough interpretation of the world, and it was not until it had been processed by the common sense that reality could set in. Reasoning, or logos, he attributed only to man, and so only man was able to infer actual events.[92] Today, we have ample evidence to the contrary showing highly intelligent behaviour in a wide range of animals, especially in vertebrates such as apes, dolphins, rats and mice. Plato was seemingly more generous than his mentor in crediting other animals with the ability to form thoughts, which made it difficult to present a clear distinction from human consciousness.[93] So if all other outward senses have an organ associated with them, what about this internal common sense? Aristotle did in fact produce a theory outlining its physiological background. Human intelligence, he explained, was located in the heart.

This quite romantic interpretation would survive for a substantial portion of human civilisation. Not until Descartes, active in France in the 17th century, would the notion be challenged to any substantial degree. As you know, this was about the same time that "internal senses" went out of fashion, and so the common sense failed to have any true renaissance despite its conceptual reworking.

Descartes quite fittingly placed the human intellect in the brain,[94] illustrated by his now iconic quote "I think, therefore I am". Finally, after several centuries, the core of human philosophy had finally found its way home. Descartes was perhaps a bit too enthusiastic though. In these times of renewed scientific pursuits, there was a certain eagerness to find logic in things. Sadly, the scientific method of empirical testing had not really been widely adopted yet, so the Common Sense was thought to be located in the pineal gland. While I do not want to unnecessarily criticise one of Europe's great thinkers, his reasoning behind this was quite out of touch even with the contemporary understandings of basic physiology.

Today, we know that the pineal gland is primarily focused on the circadian cycle; through its light-dependent production of melatonin, it regulates our hormonal levels so that we are tired during the night and alert in the morning.[1] It is roughly in the centre of the brain, and this might be an important aspect of why it was considered home to the common sense. Descartes went a bit further and claimed it was suspended in the most central ventricle, which he claimed was filled with air.[94] This was in contrast to our contemporary knowledge that the ventricles are in fact filled with liquid, which is important for draining the waste products from the brain. From a modern perspective, this is where it gets a bit strange. Descartes claimed that through the pineal gland, there is an airflow that corresponds to the human spirit. Small fibres connect it to our sensory organs. When any one of these is activated, there is a pull on the fibres causing tiny valves to open, letting air pass from the ventricles into this internal piping system. The decrease in pressure would then sensed by the pineal gland, which in turn interprets our sensory input in order to integrate it into one holistic spiritual output. While this lacks any trace of scientific truth to it, Descartes theories were important in bringing the common sense to the brain, a crucial step without which the science of neurophysiology might never have been born. Living in a time of heavy religious influence, it is not surprising that many people adopted a sort of magical thinking even during pursuits of science. Sadly, this misconception has somehow survived into our times. A quick search online will provide you with ample suggestions on how to "cleanse your pineal gland". I have even personally met a man who carried a bottle of brown water with him wherever he went, discoloured by a set of pine cones resting at the bottom. To him, the logic was quite clear: considering that pineal refers to pine cone, he staunchly argued that this liquid would not only quench his thirst but also reinvigorate his spiritual and mental faculties.

A Sense of Consciousness

As our scientific knowledge grows exponentially, we are still trying to pinpoint the point-of-origin of human consciousness. From this perspective, attributing human, or animal, intelligence to an integration of the "specific senses" into a "common sense" could have some interesting implications on a matter pondered by history's great philosophers: what governs human consciousness? Modern technology has allowed quite interesting insights on the subject, and unlike Descartes, these adopt a more scientific methodology which merits acknowledgement. The question is quite simple: are we truly in charge of our actions, or are we ruled by our sensory organs and only believe we possess a free will? I realise that such questions might be too large to answer here, and the last thing I want this book to cause is some sort of existential crisis. Studies on the field have however caused much debate. In the 1980's, American scientist Benjamin Libet pioneered the subject field. In a famous experiment, he recorded participants' brain activity during a simple task.[95] A build-up of an electrical signal prior to a muscle movement, a so-called readiness potential, had already been described in the 60's, and Libet wanted to know if this signal could be recorded before the decision to move said muscle. A build-up of energy prior to any decision would logically indicate that the brain decides on the event before the individual, or at least that was the argument. Indeed, that is exactly what happened: participants reported their intention to flick their wrist about half a second after their brain activity had shot up.

Libet's study suggested that your brain builds up energy, you decide on the action, and your body then carries out said action. Does this mean that any decision we make in response to a sensory stimulus is already predetermined based on our sensory input? Maybe. Libet's seminal work in the field of consciousness and free will has been of huge importance, but it might come as no surprise that the controversial interpretations have been challenged over time. Dr. Aaron Schurger published an article in 2012 arguing for the rise in brain activity to be considered a necessary build-up in preparation for performing the movement, and that the decision to move should be considered as an ON-switch.[96] To use an analogy to illustrate his argument, think of it like a water tap: in order for there to be running water, there must be a build-up behind the scene, and you do not turn the knob until the moment when you finally need the water. From this perspective, you can see how the build-up in energy, whether in the brain or in water, is necessary and a highly conscious preparation.

Considering that studying disorders often teaches us valuable insights into the underlying normal physiological systems, Tourette's syndrome might prove a good example in studying human consciousness and the ability to form a "common sense". Characterised by the way it causes involuntary movements or verbal statements, Tourette's has been frequently represented in popular culture. While the most common expression of this neuropsychiatric disorder is excessive blinking or involuntary grimacing, the more famous symptoms are those of coprolalia: the involuntary utterance of rude words or phrases.[97] Very little is known of the neurophysiological mechanisms behind this severely disabling disorder. Many patients might be able to withhold symptoms for a short period of time, only for the tics to manifest even more when their energy has been depleted. Some brain regions have been suggested as being more important than others in its underlying pathophysiology, with some special reference to the frontal lobe.[98] Coincidentally, this region contains the orbitofrontal cortex which is responsible for impulse control. It is also the last part of the human brain to develop under normal conditions, and it is not until the age of twenty five that the human brain is matured with the final establishment of this structure.[99] The ability to refrain from being controlled by external factors is quite clearly a late bloomer in terms of evolutionary biology, but we must accept that even the adult human is highly susceptible to the sensory systems we implement for perceiving the world outside our brains' cranial confines.

Intelligence and the concept of free will are for obvious reasons a highly debated subject, and if you find it interesting there are several books dedicated to this particular field. Many argue that questioning the concept of free will might cause universal upheavals as people stop taking responsibility for their actions, but this may suggest that free will does indeed exist, as any form of "upheaval" would involve integrating this new information about one's own consciousness in an arguably conscious manner.

Time will Tell

Having dedicated much of this book to the interpretation of the external world, let us stay a bit in the mind. Common sense is after all only one of the Five Internal Wits, which as you might remember were: common wit, imagination, fantasy, estimation and memory. Having dwelled on the common wit long enough, perhaps it would be fitting to move on to a different aspect. Out of the remaining four, I find estimation to be a rather interesting candidate. Memory deals with the past, and both imagination and fantasy arguably fabricate possible futures. Estimation instead involves taking prior experiences and extrapolating from them in an attempt to predict the future. Clearly, this is an all-important ability for any biological organism in its fight for survival. Had our hairy ancestors never learned to estimate a chain of events, we would have been a very short and bloody footnote in the annals of evolution as waltzing into the lions' den without being able to predict the outcome makes for rather poor sport.

One aspect of our estimative ability has had its importance drastically increased since the onset of the Industrial Revolution. Our ability to tell time might have been convenient during ancient times, but we never had to plan our day to the minute before the normalisation of a fixed-hour working day. Today, chronoception, the perception of the passage of time, might make or break an entire professional career. Much like the common wit, our perception of time lacks any dedicated sensory organ; there is no internal hourglass dripping organic crystals. Instead, evolution has seen it fit to allow us some physiological conceptualisation of time through the sun. This goes further than simply surveying the sun's current position. The focus returns to the pineal gland, or epiphysis, which produces melatonin and helps us sleep. Its production is inhibited by sunlight caught by our eyes, meaning that by definition there can be no such thing as a "night-person" unless they actively fight against their own biology. The circadian rhythm, or the day-night cycle, is further enforced by the hypothalamus.[19] Also influenced by the setting and rising of the sun, this part of the brain signals the adrenal glands to produce cortisol. Cortisol, in turn, is a steroid hormone that increases alertness, and having its production timed with the rising of the sun, we go through what is called the cortisol awakening response as our bodies prepare to be activated for a brand new day. Cortisol is also important in stress reactions, enabling us to quickly step up our fight-or-flight game. As a result, the impact of cortisol turns a drawn out morning procedure to a relatively fast one, and if you use an alarm clock, you probably help this stress reaction along the way with the sudden and uninviting sound it produces. Most, if not all, of us might question this from time to time, feeling like we were in fact meant to stay in bed all day. We should however remember that in our modern times with electric lighting, we are pretty good at messing with the schedule evolution set out for us.

Let us stick with the hormonal influences for a while longer. Our bodies are subject to a series of cyclical events that occur a number of times over a day-night cycle. These ultradian rhythms involve for example sleep cycles, blood pressure fluctuations and appetite. Perhaps with a few gregarious exceptions, it is rare to use one's level of hunger to tell time. Research has shown that another important hormone, dopamine, might be more involved.[100] Acting as an important excitatory neurotransmitter, the blood levels of dopamine fluctuate in 4 hour rhythms, which might explain why some are more alert at different stages of the day. While this might give us an idea of the passage of time, it can hardly be said to be useful in any real chronoceptive task- I would not trust myself to catch the train which leaves in 5 hours by making sure I am at the station in one and a half alertness-cycles. Our hormonal changes certainly help us tell time in the sense that our bodies can tell night from day, but they fall short when it comes to shorter time periods. Having grown up in Sweden, I can also vouch for the sun being a pretty useless point of reference for most of the year seeing as it never gets quite dark in the summer while the winters remain depressingly so. Regardless of our access to the sun, humans can most certainly tell time to a certain extent; for instance, if we were asked to stand up after 10 minutes, I doubt anyone would wait an hour.

For finer assessment of time, our current understanding of chronoception is that humans do possess a neural network responsible for calculating the passing of time. Functional magnetic resonance imaging (fMRI), which provides real-time imaging of any biological system, has helped gain further insights into this tricky field.[101] In one study, subjects were asked to estimate the time passed between two sound signals, while having their brains visually scrutinised for any surge in activity. Certain areas in the basal ganglia and

right temporal cortex lit up significantly more in subjects performing this task compared to those asked different questions unrelated
to time. As the parietal lobe is generally involved in integrating
sensory information and issuing motor commands, and the basal
ganglia is responsible for maintaining basic functions which make
said commands possible, it would certainly seem as if chronoception
could also biologically be justified as a type of sense. Further studies
in the field will most certainly shed further light onto this slightly
underrepresented, albeit highly useful, human ability. More than
just a matter of curiosity, this scientific pursuit is very important for
many people, as several disorders carry with them a disability to correctly assess the passage of time, such as Parkinson's disease and
attention deficit disorders like ADHD.[102] Despite fitting in with the
internal wits, perhaps we can expect chronoception to make friends
with the outer senses as time is after all an external entity, consisting
of light particles passing through space.

Friend, Foe or Self

In the example of our hairy ancestors avoiding the lions' den, perception of time is one of the factors that would have been favourable,
as a certain spot might have been safe at different times of the day,
but an equally important aspect would have been the ability to recognise the threats themselves. Similarly, we have to be able to
identify the benign, which allows us to differentiate between friendly
members of our own species and those rivals that might want to do
us harm. For this reason, familiarity has been put forward as a sensory candidate. Far from being limited to faces, individuals or even
species, familiarity refers to recognition of all things. What it boils
down to is us conceptually being aware of having performed something before, and hopefully retrieving that experience to deduce
the expected outcome of what is to happen.[103] In short, we recognise situation A, we then identify how that looked moments after in
situation B, and finally evaluate the chain-of-events that took us from

A to B. The same holds true for recognising a certain object in a familiar, or unfamiliar, setting. For instance, if you see your favourite cup on your co-worker's desk, there are two conclusions — either the sanctity of your coffee mug has been tarnished or you both shop in the same store. Based on previous encounters with your colleague, you could guess at the most likely event.

Never is our sense of familiarity stronger than during episodes of déjà vu, the strong feeling of having experienced an event before. Our current understanding is that the temporal lobe houses our ability of recognition. Memory formation happens in the temporal lobe through integrating sensory experiences in the hippocampus and amygdala[1]. On the most basic level, this allows us to respond instinctively to a threat without the information having to travel all the way to the cortex, as it would otherwise need to be for us to be made conscious of an event. The temporal cortex rests slightly above our ears, and studies in rats have shown that it is specifically the medial area that is responsible for our ability to sense familiarity. An activation of this area, the perirhinal cortex to be exact, is strongly associated with viewing familiar objects.[104] Damages in this spot have produced rats who treat even familiar objects as if they were seeing them for the first time. Altogether, it would seem as if familiarity, much like chronoception, therefore has both a physiological and perceptual claim to being elevated to the status of an official sense.

Simply sensing that time has passed, and recognising that the animal charging at you with its claws out might not be very friendly, is however not going to get anyone very far up the evolutionary tree. A puma needs to find the exact right time to strike and the llama must counter every move with such precision that the slightest mistake might mean its demise. This leads us to the final candidate in this chapter on sensory candidates: the sense of agency. Agency refers to the perceptual feeling of owning oneself, and of having selected to perform a certain action.[105] For instance, when you cross the street, you do not simply respond to sensory input as it comes, but rather you take in all the sights and sounds of your surroundings and make a conscious decision on when to make your move. Our sense of agency is one of the strongest arguments for us possessing a free will rather than being moved by sensory chaos. One could argue that it was this particular sense that the Libet experiment dealt with, with the delay between a surge in brain activity and the vocalised intention to perform an action indicating the time needed for our human perception of agency to manifest itself.

This internal expression of our free will is not limited to our motor actions. Speaking is an example of a complicated mixture of internal thoughts made audible by a set of motor commands. Forming words necessitates an active decision, but before that, the words themselves have gone through some internal revision so they can emerge in a way that makes sense. We rarely weigh each word on a golden internal scale before producing them as sounds, but nevertheless, there are few things that make us stand out as individuals more than the words we speak, reflecting the inner workings of our minds. Most of our individuality is after all happening in the depths of the brain, and so it is perhaps here that our sense of agency is the most important. Similar to examples given in previous chapters, study of certain disorders once again sheds light on the significance of normal physiological mechanisms. For instance, patients suffering from psychotic symptoms, such as those seen in schizophrenia, frequently describe how they feel controlled by an outside entity, as if someone else is in control of their thoughts.[105] Hearing voices in your head is perhaps the most striking example of these symptoms, with the thoughts being there, but as the patient does not feel in charge of them, they are manifest as external voices, often of a very disturbing nature. Experiments have shown that a sense of agency can also be induced by outside forces. In 1999, the University of Virginia published a study by Wegner and Wheatley which concluded that participants could be made to believe they were in charge of an action despite someone else performing it.[106] Subjects were made to move a mouse which was simultaneously controlled by a second individual. Despite this second participant being able to stop the mouse at certain intervals, the primary test subjects still reported it as their intention to perform the action.

Solving the puzzling biology behind our sense of agency would be a key step in cracking the question of human consciousness. At present, our knowledge in this field is rather limited, but once again the temporal lobe seems to be a key player, especially in its connection to the parietal lobe where our motor and sensory input is managed.[107] While this might appear somewhat vague, the main point is that agency is seemingly something which is biologically supported through our many neural networks, and as such can be mapped out and treated. Perhaps our sense of agency also falls into the concept of a common sense as explained by Aristotle, with the many specific senses integrated into one consciousness. Patients with damage to the junction between the temporal and parietal lobes frequently exhibit symptoms associated with a lack of conscious self-control. This ranges from asomatognosia, characterised by the inability to accept that certain body parts belong to them,[108] to anosognosia, where patients still believe they can move a completely paralyzed body part despite sensory protests telling them otherwise.[109] If you are as intrigued by these phenomena as I am, I would strongly encourage you to read the fascinating accounts of cases such as these by writers such as Oliver Sacks. Humans are of course not alone in having a sense of agency, but investigating them without access to a mutual and nuanced language might be difficult.

What It Means to Sense

As we stated at the start of this chapter, there are certainly a number of purely physiological methods of perception that one could argue deserve the status of sense. In addition the candidates discussed above, our ability to detect oxygen is a further example and allows us to leave the increasingly philosophical area of internal wits behind us. Being able to detect the substance that fuels all of our internal processes is obviously quite important and we have specific receptors dedicated to this particular task. So called chemoreceptors, detecting chemical compounds, are found in strategic places inside our blood stream. Some of them certainly do detect oxygen, but the ones most important for our automated breathing are those detecting carbon dioxide, as these are the ones that dictate our respiratory rate.[110] Carbon dioxide is merely a by-product from the cells. Produced as our cells convert oxygen into energy, carbon dioxide can be used as an indirect measurement of how much re-fuelling our lungs need to do, with a high level of carbon dioxide suggesting the cells are hard at work and that we need to breathe faster to prevent the acidity of the chemicals affecting our blood and organ function.

From this perspective, chemoreception, specifically the perception of carbon dioxide and oxygen, could well be considered a physiological sense. Where would it fit though? It is not necessarily an internal wit, since it deals with the physical world, but neither is it telling us much about the external environment. We could argue that a mountain climber can detect oxygen levels by the increasing shortness of breath, but breathlessness is only a symptom. If a sense is meant to give us a way of interpreting the world, one can argue it must also produce a recognisable and conscious response. Yet, important as it is, the detection of certain gases is completely internal and only for the benefit of subconscious regions in the brain which then drive homeostasis, rather than it being something we are conscious of doing.

The list of possible senses can therefore be made much, much longer. Balance, pain, temperature, the common sense, time, familiarity and agency are examples that are justifiable additions to the sensory pantheon. Having been given little love over the last centuries, I especially find the inner senses deserving of more attention, and while being more difficult to pin down physiologically, most of us would probably agree on their importance in how we make sense of things. I dare not say "all", because our perception of the world is undoubtedly highly subjective and the inner wits sort of fall apart if you do not believe in free will. Sure, as a species we all agree that a cactus offers less in the way of inviting tactile stimulation than a fluffy cat, but more nuanced differences in sensory perception. A fire alarm seems to pierce the soul simply because their inventors had human hearing and designed the sound to be as noticeable as possible. In this regard, the world appears distinctively the same no matter where in the world you find yourself. We are after all human, and any further interpretation of these specific sensory inputs is quite small in the grander scheme of things. Still, those deviations are what make us individuals and no existing sense can explain how this comes about. In order to create our own subjective perception of the world, we need to accept the existence of internal senses. Perhaps acknowledging this phenomenon is a step towards a more accepting society; we would not shun someone for not enjoying a specific art-form due to their visual preferences, so why alienate ourselves based on the internal perceptions of things. As a species, we perceive our surroundings virtually identically, but as individuals we need to dig deeper into the internal methods by which we assess the world.

It is somewhat difficult discussing the concept of internal senses without some philosophical meandering, but it is certainly true that in order to take in the finer aspects of the world, we need to first look within ourselves and identify those prejudices that we all undoubtedly have. Hopefully we can change some of them, but if not, we can at least accept that they are there, and understand why others might not agree with subjects that to us seem obvious. After all, we would never claim that a person with glasses is incapable of enjoying an art gallery, nor would we berate an old man with a hearing aid for trying to enjoy the music of his generation. Our senses are fallible. Just as it would be unwise to drive a car without prescription glasses, perhaps we would do well to accept that our personal interpretation of an event may occasionally be objectively incorrect. In the same vein, you may not agree with the list of extra senses I have discussed in this chapter. I nevertheless hope we can agree that there are more methods of interpreting the world than our popularised five senses.

Chapter 8

Making Sense Across the Animal World

Most of us are aware that our human senses are far from the only approaches to perceiving the world. Indeed, the English language has adopted several idioms that allude to more sophisticated senses found in the animal kingdom. You may have come across expressions such as having the "eyes of a hawk", meaning you are unlikely to require prescription glasses, while someone described as a "bloodhound" will not rest until their target is sniffed out. In the next chapter, we will deal with some of the senses that evolution has deemed off-limits for mankind, though they are available to some of the non-human inhabitants of this planet. While the sensitivity of some animal senses far exceeds human capabilities, such as a dolphin's impressive hearing and a shark's sharp nose for smells, we should remember that we are all walking the slow path of evolution, and what is considered superior in one environment might be detrimental in another.

The purpose of this chapter is to introduce some of the many ways matters of sight, hearing, smell, taste and touch have been dealt with across our planet. It would not be possible to do each system justice in its entirety, lest this book be five times longer. This overview therefore should act as an invitation to learn more should you come across a phenomenon that stands out. Let us focus on some notable examples of sensory prowess, investigate the most notable way in which they stand out, and why Nature has made it so. By looking at life from the perspective of those around us, perhaps we may even learn something about ourselves.

Eyes in the Animal Kingdom

As you will see in the next chapter, Nature has provided Earth's inhabitants with an array of solutions for translating light into visual images. Even though some eyes might look similar at a glance, we can find slight differences, for better or worse, hidden behind the lens. The eyes are excellent illustrators of the concept of convergent evolution, where nature has found different but similar solutions to the same problem. The octopus, for example, sports what is referred to as cephalopod eyes, which differ in certain key areas compared to our vertebrate variety.[111] While the optic nerve punches a hole in the human retina to reach the rods and cones, the octopus' optic nerve simply approaches the retina from behind. This means that there are no nerve fibres climbing the retina, blocking incoming light and creating a blind spot. Instead of a small fovea to provide adequate visual acuity, the octopus' eye therefore receives visual input equally from every direction. The separate evolutionary trajectories that produced human and octopi eyes represents convergent evolution in action, or different means of natural selection leading to the same general outcome. As you will soon see, there are many more alternatives to choose from when perusing the smorgasbord of vision. Let us therefore dig a little bit deeper, and look into the eyes of some of our Earthly cohabitants.

Different Answers to the Same Question

Given Earth's vastly different environments capable of supporting life, it may not come as much of a surprise that animals have developed different eyes and different approaches to vision, depending on where evolution has taken them. As we mentioned before, the fact that multiple variations of eyes have developed independently of each other is called *convergent evolution.*

On a molecular level, vision predates all forms of eyes, albeit not image-forming vision. I say this because all eyes, regardless of their owners, are built up of photoreceptors that are very similar at the amino acid level, suggesting a single ancient origin.[112] Our different environments subsequently altered the way in which the visual network was built up. Evidence of this can be seen in how the neurons that receive signals from the photoreceptors are distributed in the retina. These are called *ganglion cells,* and an animal living in flat terrain may have cells gathered in "visual streaks", corresponding to the horizon of the field in which it lives.[113] For instance, this is true for both rabbits and cheetahs. Species that live in dense vegetation, including humans, instead have their retinal cells in a more circular and symmetrical pattern, which allows more lateral and peripheral viewing.[19,113] This highlights how our anatomy reflects our surroundings and also influences the very way we see it as ganglion cells collect information from the photoreceptors and package them before sending them on to the brain, with their distribution consequently guiding our perception of a scene.

While notable, this structural difference is quite small, near insignificant, compared to those found between animal phyla. The structural anatomy of the eye itself may reveal plenty of information of how its owner uses vision. When reading this chapter on animal vision, it might be helpful to return to Figure 7 from time to time to get an idea of how light actually enters each eye.

A Master of Vision

While no eye can be considered superior, since it is designed for its specific environment, it is tempting to highlight an animal whose looks reflect the vast array of visual information it can process, as this sensory specialist's carapace ranges from calm green nuances to neon blue and psychedelic pink. The Mantis shrimp is certainly a master of all things visual, and it looks the part.[114]

·········· Incoming light is transformed into neural impulses through a multitude of different eyes

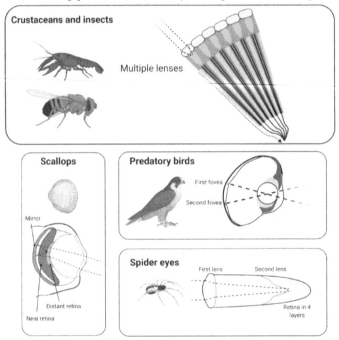

Figure 7.   Animal vision. A short exposé of animal vision and eye types. At the top is a photograph of the colourful mantis shrimp, featuring two compound eyes. Eye like these operate according to the same principles illustrated below for crustaceans and insects. The other boxes show how light (dotted line) enters the eyes of different animals before reaching the retina. Created using Biorender. The photograph of the mantis shrimp was taken by prilfish and is shared in accordance with Creative Commons (CC BY 2.0).

Ironically, the only conventional thing about the Mantis shrimp may be its retinas, of which it has two, one for each eye. These are capable of detecting light in two dimensions just like us. When you look closer, however, you will see strange patterns crossing the eyes. A strip of six rows passes through the centre, made up of smaller compound eyes called *ommatidia*. Within these eyes, we find not three types of colour receptors as in humans, but an astonishing 12, enabling the Mantis shrimp's compound eyes to view light on a spectrum ranging from deep ultraviolet to intense red.

Most scientists agree that three or four types of photoreceptors are enough for distinguishing between colours, with some animals having up to eight in cases where they might need extreme colour acuity. So what has brought upon this crustacean hubris? The answer lies in the way the Mantis shrimp uses its eyes to detect colour. Essentially, the object being viewed must be centred on one of its bands of compound eyes. The Mantis shrimp will then move the eye across the object, effectively scanning it like a barcode. This type of scanning suggests the Mantis shrimp is only able to recognise colours individually rather than actually discriminating between them. If this is true, this would essentially mean that the viewing conditions lead to the animal's individual eyes performing rather poorly and that its extra photoreceptors have evolved as a means to counter this. Essentially, the mantis shrimp may scan a colourful scarf on the ocean floor and when it reaches a break in the colour code, it will simply think "Ah, yes, here is a colour!" However, it will probably not be able to determine that it is a significantly different colour. While this way of thinking about colours might seem pretty strange to our version of vision, the mantis shrimp certainly stands out in its vibrant shell and sensitive eyes.

Satellite Dishes and Seeing in the Dark

Not far from the mantis shrimp, we might find a neighbour with a different solution to the puzzle of sight. The scallop, a type of salt-water clam, is a common sight in all oceans across the planet. Like its other mollusc-relatives, the scallop makes use of pinhole eyes rather than our bulbous counterparts. Nevertheless, these pinholes lead to a unique system for processing visual information.[115] At the end of each scallop's pinhole eye, there are a series of concave reflectors consisting of tapetum, a bio-reflective material also found in cats' eyes which gives them excellent night vision, not to mention eyes that glow in the dark. This concave "dish" redirects the light onto a two-layered retina. Depending on the angle of the light being reflected, the beam will fall on different parts of this retina, just like it may do in humans. In scallops, when the light falls on a distant layer of the retina, away from the satellite-like dish, the photoreceptors will give an off-response, indicating that the image or the object being viewed is not straight in front of it.[115] As the image moves in front of the scallop's many pinhole eyes, it will cross successive receptors, telling their owner that whatever is presenting the image is on the move. When the image has moved over enough receptors, the scallop will simply close, protecting itself from the approaching predator in its hard shell. The use of concave mirrors to focus a signal is something we have adopted in the design of satellite dishes. This is just one of many, many objects of human design which has a conceptual counterpart that can be found in Nature. Considering the eons that have been available to Mother Nature, I think we can see these occurrences as a figurative thumbs-up to its clever design.

Cats and scallops have completely different evolutionary pathways, and even eyes, but still share the use of tapetum as a means of surveying the dimly lit corners of our planet. While cat eyes are mammalian in nature, and resemble ours far more than those of parabolic antennas, the way in which a particular mechanism is similarly used in two divergent species invites some further comments, particularly on feline vision and their slit pupils. These narrow passages are however not unique to cats, as you can see them in other predators such as crocodiles and snakes. Whether reptile or mammal, the purpose is the same: allowing precise control of just how much light enters the eye.[116] Humans can change their pupil-size about 15-fold from smallest to biggest, which happens automatically in response to light conditions, with dimmer conditions resulting in increased pupil diameter so that more light may enter. However, this is far inferior relative to that seen in cats, whereby their slit pupils can change 300 fold of its extreme values, reducing glare on sunny days and increasing visibility at night. This is a valuable example of how a creature's appearance, in this case cats' slit pupils, reveals their innate behaviours. Some may argue differently, but humans are clearly not geared towards being night-owls, which we can tell from the way Nature has moulded our ocular organs.

A Third Eye

Taking a couple of steps up the evolutionary ladder, we find ourselves amongst the reptilians, leisurely heating their cold-blooded selves underneath the sun. While these scaly lizards have little in common with the original organism capable of sight, they have retained a key characteristic, namely a third eye which is also called the parietal or pineal eye.[36] This eye seems to have originally functioned as an alarm bell for any approaching dangers from above. Lizards, however, are generally rather safe from aquatic attacks, thanks to their evolutionary decision to grow legs and move up the beach. Instead, their parietal eye mainly deals with regulating temperature, as it determines heat through calculating the finer aspects of sunlight.

This type of third eye has taken an interesting path throughout evolution and you can even argue that we humans retain it in a way. Instead of a pineal eye, we have a pineal gland.[37] While the pineal, or parietal, eye is a type of photoreceptor located on the top of a lizard's head, the human pineal gland is safely tucked away close to the brainstem. Even so, this gland is an atrophied version of its seeing counterpart, having lost its visual prowess somewhere along the winding road of primate evolution. Though it does not detect the sun directly, it still plays a major role in helping us adapt to light and darkness. The pineal gland, hidden in the dark underbelly of the brain, still receives input from the eyes. When no light hits the retinas, such as during the night, the gland produces melatonin, a hormone telling us that it is time to go to sleep. During daylight hours, the signal from the eyes instead triggers the pineal gland to break down the melatonin, making us more alert. In this way, the human parietal eye helps us retain a day-and-night cycle.

Eyes Plentiful

Another animal that has got eyes specifically adapted for the setting of the sun is the fly. While a fly might not be much to look *at*, the fly itself has a lot to look *with*. These flying insects have eyes set in what is known as *neural superposition*.[117] What this means is that the light sensitive areas are divided into distinct divisions, so rather than having one big retina doing the job like in humans, the fly eye works through team effort. Visual data is combined with that of the neighbouring areas and relayed to the brain by immaculate neural wiring. This is done by having a visual image registered separately by each eye, where it is portrayed upside down just like in humans. After this first step, the signal is relayed to the same brain region as the adjacent eyes.[117] In this way, the effective size of the original signal is increased sevenfold, making it a classic example of how the sum of an object can be much greater than its parts. With regards to flies having adapted to the setting of the sun, this can be translated as their eyes allowing them fifteen more minutes of useful vision at sunset and dawn.[118] The design of the visual pathway in flies allows them to have both a very high resolution of an image as well as an incredible sensitivity to movements. They are particularly sensitive to the latter, and this can be attributed to the use of their neutrally super-positioned eyes, acting as highly sensitive elementary motion detectors which relay spatial information to a central integration point.[119] These individual motion detectors are further interconnected in a complex network, creating a mapping system where even the smallest of movements can be detected. Once again, we find that the process of convergent evolution has produced several solutions to the same problems, even across such vast evolutionary distances as between mammal and fly.

I thought we would continue this section on strange animal eyes with some that many might find a bit unsettling, namely that of spiders. While most people imagine them to have a myriad of compound eyes, like those of the fly, most spiders actually have eight eyes, with some sporting only a modest six or even fewer.[120] Running eight eyes takes energy and some arachnid species have simply dropped a few as they have adapted to their niche habitats. Out of these eyes, there are actually only two main ones actively looking, and in many ways they are more similar to our human version than to those of the fly. The remaining eyes sort of exist in position, absorbing the surroundings. You could liken the spider's eyes to a bunch of CCTV cameras, with two main ones keeping active lookout for prey, and a couple of stationary ones surrounding them in a fixed position as a means of widening the periphery. This means that the two central eyes provide the stereoscopic vision which visually-guided predators rely heavily upon, with the remaining stationary counterparts being more sensitive to motion.

While most are quite ocularly gifted, some spiders might however be more equal than others in terms of their visual prowess. Jumping spiders, sporting 4 eyes, have very complex retinas, divided into multiple layers.[121] These retinas are composed of 5–7 photoreceptors in width and 50 in length, and can be moved and even rotated by the main two eyes. In contrast, the other stationary eyes have fixed two-dimensional retinas which simply serve to detect movement and act as pathfinders for the two moveable, principal eyes.

Predatory Birds' Need for Accuracy

Let us end this chapter by looking up into the sky. Birds naturally need to have excellent vision if they are to spot things from their cruising altitudes, otherwise a myopic eagle would probably stay hungry for a very long time. I am going to start with a nocturnal animal, well known for its large eyes and academic predisposition: the owl. Notably, owls do not have eyeballs in the traditional sense.[122] In fact, they have ocular tubes stretching into their eye sockets until the skull closes in on them, holding them in place. As a result, owls cannot move their eyes, instead opting to simply rotate their entire heads up to 270°.[122] Humans, on the other hand, can only reach a visual field of 180° despite our well-developed eye muscles. We previously talked about the two photoreceptors, rods and cones, with rods being better suited for darkness while cones can detect colour. It might not come as a surprise then that owls have an impressive concentration of rods all over their retina, and very few cones. The owl world is consequently very monochrome, but it does provide them with superhuman night vision. Their pupils are also highly sensitive to light and can adapt to most conditions, which means that even the brightness of day will not appear blinding, even though the owl will most likely be sleeping then anyway. Sadly, the owl, favouring visual acuity at a distance, suffers from severe hyperopia and are essentially born with a need for reading glasses. Lacking any natural optometrists, the owl has to settle for a more tactile approach, using tiny whiskers to gather information on things right in front of it.[122] They do take good care of their eyes though, possessing three sets of eyelids; the upper one blinks, the lower one blocks out the sunlight during sleep, and the third one shuts diagonally across the eye. Thanks to the latter being translucent, the owl can use its eyelids like a set of biological safety goggles when swooping down through the branches. This somewhat divergent solution to the visual problem highlights an important aspect of sensory neuroscience: sensory trade-offs. The owl, often preferring dense and dim-lit foliage for its hunting ground, has excellent hearing. We will not go deeper into that here, but it may be interesting to point out that owls appear to have a relatively smaller region in the brain dedicated to vision relative to hearing, allowing auditory-guided prey localisation.[123] Such trade-offs will become even more apparent when we discuss the intricate sensory arms race between bats and moths later in the hearing chapter.

Clearly the owl has evolved with a respectable reverence for the visual system and highly specialised visual acuity. Still, a popular idiom referring to visual prowess favours another avian creature. In that regard, not even owls *have eyes like a hawk*. Unlike the owl, the hawk is not necessarily very keen on hunting in the dark. Its retina reflects this by having a high density of colour-coding cones at its disposal.[124] Not only that, but unlike humans or even owls, hawks possess two foveas; one is central and positioned much like ours, the other is temporal and hit by light coming across the beak.[124] In reality, the central one is deeper and produces an image with better resolution than the temporal shallow one, but they both still contribute to the hawk's amazing visual acuity. This feature is not exclusive to hawks however but is seen in other predatory birds as well, notably in eagles and falcons. But is there a downside to having two foveas? As mentioned earlier, the fovea is meant to create a sharp image of *what* is being seen. The rods, on the other hand, are much better at helping us detect motion, relaying *where* something is going. Humans have spent a significantly longer time as prey than predator, and as a result, we have always needed to stay on top of things moving in our periphery, prioritising motion. Hawks, on the other hand, have no predators to worry about and therefore have no real need to run or hide when confronted with visual motion. Therefore, they need precise visual cues on how to chase down their targets, so a hawk arguably has the genetically predefined eyes of a killer.

The visual specialisation does not end there; while we humans regularly accommodate by changing the shape of our lens, many birds have additional muscles allowing them to alter their eyes.[125] Crampton's muscles, which exist in birds as docile as the chicken, stem from the anterior segment of the ciliary muscles that control the lens and attach to the cornea. This extra addition allows much greater precision, and scope, for focusing in on a target. Here we have an example of something very familiar to us, accommodation, but implemented in a way that is completely foreign to our species; considering the importance we humans place on vision, it is interesting to wonder why such a mutation did not emerge somewhere in our evolutionary past. On top of this advantage in accommodation, not only do birds see things clearer, they also see things *faster*.

Flicker fusion frequencies are used to measure how quickly someone can differentiate between light and darkness. If you were to see a flicker light, for example, that pulsation would slowly transform into a steady light source as its frequency increased. Clinically, these measurements can be used to quantify patients' sensitivity to visual motion, but it also tells us something about our species' visual prowess. Humans are actually quite good at detecting visual motion, and if we were to compare our eyes to cameras, you could say that we record things at framerates of 50–100 Hz. Many small birds sport much more sensitive hardware however, capable of detecting movement at 130–146 Hz.[126] In real terms, this means that birds can see things that may appear invisible to us as we are incapable of updating our visual input fast enough to conceptualise what is happening. Notably, this is in stark contrast to predatory birds, who sacrifice this temporal resolution in favour of visual acuity. Why this is so is reflected in their respective habitats, with small birds aiming to capture small insects which buzz quickly through dense foliage, while predatory birds such as the eagle hunt across open plains. As a result, the eyes of smaller birds have favoured higher refresh rates of their neural signalling, with ensuing loss of visual acuity; the same trade-off is made in your camera every time you adjust the framerate or resolution.

Of Snakes and Men

Thus far, we have largely focused on animals who experience the world very differently from us humans. Making a swift shift onto more familiar ground and looking into the eyes of our monkey cousins, their neurophysiology is much the same as ours, but there are some interesting and rather unique aspects of the primate visual development that deserve some mention. Have you ever walked down a country road, when suddenly something long and dark slithers by on the ground? Before you can realise what it is, your brainstem takes over and stops you dead in your tracks. Just as your fight-or-flight reflex kicks in, you realise that the thing you mistook for a snake was in fact nothing but a harmless stick. In fact, it is believed that primates evolved in such close relationship with the dangers of snakes that our visual system has developed a mechanism for detecting them. This *Snake Detection Theory* suggests a snake-detection pathway was hard-wired into our brains[46] and may even explain why ophidiophobia, the fear of snakes, is the most common phobia with one third of humans experiencing it.[47]

Moving back to apes, the importance of snake-detection is physiologically reflected in the disproportionally large region of the primate hypothalamus called the pulvinar.[48] Although the extent to which the Snake Detection Theory holds true continues to be debated, we do know that there is a neural connection between the superior colliculus, pulvinar, and amygdala.[49] As we discussed earlier, the superior colliculus is a key junction in relaying visual information and the amygdala is the region of the brain dealing with fear. Rather than forming a visual representation of the light hitting the retina, it is argued that this pathway has a fixed set of responses, increasing reflexive visual alertness in the face of danger. From the perspective of evolution, this makes a lot of sense, since it allows us to respond to a threat even before realising what it is; mistaking branches for snakes is a small price to pay in the greater scheme of things.

Such hardwired subconscious responses are not only limited to snake-detection. Phenomeona such as blindsight, whereby an individual can still respond to certain alarming visuals through subcortical pathways; these neural arcs may allow someone with damages to the occipital cortex the chance of catching a ball being thrown at them even though they cannot actually point to or describe the approaching object.[50] Blind-sight certainly indicates that consciously understanding what we see might not necessarily be the same as not possessing vision. There certainly are other interesting tricks played by brains on the visual systems across the animal kingdom. Tiger beetles are for example rendered blind as they chase down prey[51] as their brain cannot keep up with the processing power needed to integrate the visual information that is fast incoming through their speedy movement. As a result, the beetle has to make several small stops during a hunt and recalibrate its attack trajectory each time.

Returning to snake-human interactions, it may be fitting to view things from the reptilian perspective as well. When introducing human vision, we covered the misconceptions of IR and UV radiation wrongly being labelled as light. Had humans evolved from snakes rather than primates, this distinction would likely have looked very different. Snakes walk the tight-rope of what we could call vision and a separate IR-detecting sense. Derived from cells sensitive to touch, or more specifically heat, pit organs are embedded in the snake's snout.[52] Picking up on wavelengths invisible to us humans, these allow their owner to create heat-maps of their prey. Incoming photons are directed towards thin membranes at the bottom of each pit, which is surrounded by air pockets to minimise heat loss. IR-sensitive portions of the trigeminal nerve connect to these membranes to pass the signal on to the brain, where it interconnects with several structures that merge light- and infrared detecting systems. Whether pit organs may just be called eyes is a matter for debate as they aid in image-forming vision, but it just happens that the radiation they detect lies outside the visible spectrum for us humans. While such a debate may have little influence on either human or snake lifestyles, it offers an interesting perspective on how we view the senses.

That about wraps it up for our little exposé on animal vision. So important is vision that Nature has found several ways of producing the eye: from the pinholes of the scallop to the double fovea of the falcon, our eyes reflect the breadth of Earth's ecosystems. Still, not all corners of the world are reached by light. In these cases, other senses often need to take over, so perhaps we should be a little more hesitant before we crown vision as the king of all senses. After all, the world is a large place and there is certainly more out there than meets the eye.

Ears in the Animal Kingdom

When looking at the ears of the animal world, we need to lay down some boundaries of what constitutes hearing. Some cases, such as cats and dogs who sport very similar ears to us, may appear quite straight forward as the principles governing their perception of sound are largely the same as ours. But what about animals lacking external ears? Snakes for example do have internal ears, well-suited for discerning vibrations through the ground, but the absence of external- or middle ear structures renders them deaf to the alluring tunes of a snake charmer.[127] On the other end, dolphins also lack external ears but have such excellent command of hearing that they use it much like we would use our vision. We saw in the chapter on touch that we have receptors dedicated to vibrations, so it might be tempting to call a snake's perception of sound a variation on the theme of touch rather than hearing as the vibrations are carried through the ground. Other questions include whether we can call it hearing irrespective of whether the waves travel through gases or liquids. As we explore the best hearers in the world, it may be interesting to keep questions like these in mind. The ears of animals may look different from one another, but once again Nature has offered them all different solutions on how to harness sound.

Seeing with Your Ears

Only after 200,000 years of progress did our species develop the radar system, capable of mimicking the innate capacities of echolocating

animals. In contrast to humans, some animals are handed echolocation on a silver platter. *Blind as a bat* is a popular English idiom, but
this simile is both unfair and untrue. First of all, there are about 1,300
species of bats, and secondly, none of them are blind.[128–130] While
some of them may not have the best visual acuity, what they lack in
vision they more than make up for by warping soundwaves to suit
their needs. Our human non-biological version of this is sonar, which
stands for sound-navigation-and-ranging. Some bats produce the outgoing soundwaves by clicking their tongues or contracting their
larynx, essentially mimicking a finger-snap with their vocal cords.[130]
While most echolocating bats produce the sound through their
mouths, some do it through horseshoe or leaf-shaped noses acting
like megaphones.

While humans can hear up to 20,000 Hz, bats' hearing reaches
ten times that number at 200,000 Hz, although not all of them are
inaudible to us. If you find yourself in a dark cave and you hear the
sound of small pebbles being hit together flying to-and-fro, you
might well be hearing bat echolocation in action. Different circumstances call for different types of sounds; high frequencies, with
extremely speedy vibrations, reveal fast changes in the surrounding
areas, but they do not travel very far. As a result, most bats rely on
this higher frequency when hunting. When they need to orientate
in a larger area, some are instead capable of producing lower frequencies. To put this into perspective, echolocation is typically only
effective up to 17 metres if the bat wants to get a rough idea of the
neighbourhood, while the speedier update frequency becomes
unreliable even at 85 cm.[131] Beyond that, even echolocating bats
need vision to get an idea of what the world looks like. As their vision
is not very good, bats mostly use it to discern big changes in their
environment, such as daybreak and the setting of the sun.

There are some interesting similarities between human eyes and
bat ears. You might recall the human fovea, where human vision is
the sharpest as it is hit by focused light. Some bat ears interestingly
have an *acoustic fovea*.[132] The eardrum of the Greater Horseshoe bat

is not symmetric in thickness or width, forming a long and thick area that resonates with the frequency produced by the bat's call. As such, this area has developed a very narrow range of pitch, making it similar to a fovea in that it focuses soundwaves.

Even though we do not hear them, bat calls are by no means silent as their clicking ranges from 50–120 dB.[133] As our hair cells fail to pick up these high frequencies, it does not matter how high they go as we will forever remain none-the-wiser. The bats are not as lucky. Just because they can produce the sounds, this does not mean their hearing cannot be harmed by the high clicks. Bats do however have a pretty ingenious safety mechanism: the muscle that contracts their inner ear bones contracts just before they emit their sound, freezing the conduction of any vibrations. So while the ear might pick up the sound, the energy will never reach the hair cells. As the muscle relaxes immediately afterwards, the ear is then free to convey the click's echo which hopefully has bounced off a nearby object such as a fly, signalling that dinner is served.

The bats' use of echolocation may inadvertently have caused an evolutionary champion in its main adversary. In order to stay out of the way of these flying hunters, the moth was forced to become the best hearer in the world. Hearing frequencies of up to 300,000 Hz, these insects can detect changes in their surroundings at an amazing speed.[134] With their eardrums being sensitive to movements the size of 140 pictometres, understanding the impact that has on their hearing is a daunting challenge.[135] Think of it like this: making a moth take notice of you is as difficult as moving an atom. Having such sensitive hearing might sound more like a curse than an advantage, but studies have shown that moth physiology allows them to filter background noises of lower pitch. As a consequence, moths are essentially rendered deaf to lower notes, which makes sense as the purpose is to detect incoming bats which use high-frequency clicks themselves. The moth therefore went ahead and simply developed better hearing in the evolutionary arms-race of the senses.

There are even deadlier predators than the bat which can sense vibrations not only through water but even muscle. These swift killers can pinpoint organs, avoid bones and hone in on a target's most vulnerable areas. As if that was not enough, these apex predators are one of the most intelligent species on earth. Luckily for us, dolphins seem to treat us humans more like playmates above anything else. These cetaceans produce high-frequency clicks that are emitted through their nasal sacs and amplified through their large heads.[131] The amplification occurs through the dolphin foreheads called *melons*, which are filled with fatty tissue that channels the outgoing vibrations. The echoes are then picked up through the dolphin lower jaw. One might recall the origins of hearing here, and how the lower jaws of fish gradually evolved into the small bones of the inner ear in us humans. Dolphins have retained their lower jaws as vibration-sensors, focusing the energy towards their ears as seen in Figure 8.

There is still much to learn about dolphin echolocation. We do know it only works in water and as soundwaves move much faster in water than in air, 1,530 meters/second compared to 340 respectively, this makes their sound-based orientation much more effective than that of bats.[136] Interestingly, as humans are 60% water, this allows dolphins to figuratively see right through us, even allowing them to detect our internal organs through echolocation. This naturally makes them amazing hunters, essentially capable of biological X-ray vision.

Hearing with Your Feet

Even though dolphins pick up soundwaves with their jaw, the sound is still converted to electrical impulses in the inner ear just as in humans. While this might seem like a bit of a distance, having a soundwave travel through the head, there exists another hearer out there that allows the vibrations to travel an even greater bone-lead route to enter its oversized ears. Vibrations picked up through the toenails, sensed by sensitive nerve endings in the feet, travel through their dense bones all the way through the skull to the inner ears.[81] Unsurprisingly, elephants have excellent normal hearing, as the size

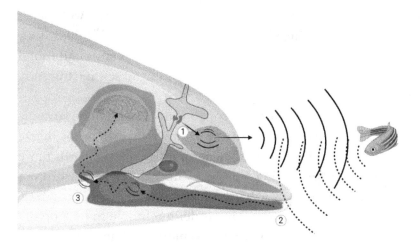

1. A click is created in the blowhole by phonic lips, causing vibrations to travel through the melon where they are reinforced before entering the surrounding water

2. Vibrations reflected by a target reach the mandible

3. An acoustic window enhances the soundwaves before they reach the inner ear, which once more amplifies the vibrations before sending them on to the brain as neural impulses to be interpreted as sound

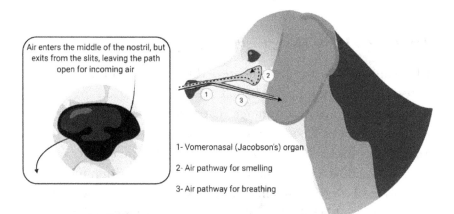

Air enters the middle of the nostril, but exits from the slits, leaving the path open for incoming air

1- Vomeronasal (Jacobson's) organ

2- Air pathway for smelling

3- Air pathway for breathing

Figure 8. Animal hearing and smell. The top figure shows echolocation in the dolphin, as they make use of sound waves to survey their surroundings. The bottom illustration outlines the dog's anatomy for smelling. Dogs' sense of smell is aided by the anatomy of their airways. Air will enter the nose and quickly reach the vomero-nasal organ to allow a pheromonal response, before being separated into two pathways. Some air will be trapped in such a way that smell molecules may have a greater chance of reaching their receptors, while a direct pathway to the lungs is used for breathing. Figure created using Biorender.

of their ears might imply, and they have been credited with being able to hear approaching storms as far as 240 km away.[82] Still, a study carried out in Zimbabwe monitored a herd of elephants and found that they would detect earthborn soundwaves from seismic transmitters up to 32 km from the source.[83]

Let us consider *why* hearing through your legs might be advantageous. Firstly, elephants are pretty much unrivalled in terms of size on land and when they move, they naturally make the earth move underneath their feet. Being able to identify other herds, or even stray members of your own herd, clearly has evolutionary advantages. While their excellent hearing is definitely able to pick up the tooting of a distant trunk, there is also merit to detecting groundwaves. Actually, it has been estimated that airborne communication falls short compared to its seismic counterpart, being perceivable by elephants at just under 10 km, a third of the ground-borne distance.[83] Being able to communicate discreetly several kilometres away, these giants are essentially part of a secret underground texting group.

Hearing Low While Flying High

If bats and moths are great hearers due to their ability to pick up ultrasound, one might think that an animal specialising in infrasound might be at a disadvantage. The truth is that the best navigator in the world uses frequencies as low as 0.05 Hz in its orientation around the globe.[84] The homing pigeon's immaculate sense of direction made it a common method of sending written messages in the past and their reliability in always finding their way "home" made them highly valuable soldiers on the battlefields through several wars, with them being famously on the frontlines of World War One.

The reason for pigeons' impressive sense of direction is however not known. Theories put forward involve them being able to detect the planet's magnetic field, or carrying iron ions embedded in their beaks that allow them to act as in-built compasses. Through sensing low frequency sounds, it is also believed that homing pigeons can detect air turbulences, finding their way safely through the air.[84] While high frequencies may produce a detailed map of the immediate surroundings, the low frequencies detected by the homing pigeon might not give it much information about the surrounding area but it will tell a story of the weather conditions far, far away. As a comparison, there are several human outposts constructed with the main focus of picking up infrasound. This was done as a response to the Comprehensive Nuclear Test-Ban Treaty (CTBT), with the equipment sensing any changes in air pressure that could be due to an illegal nuclear detonation.[85] These stations implement a frequency of 20 Hz which gives them a rather short scanning range of sensing air turbulences of up to 9 km. Lower frequencies cannot be used as our technology cannot interpret the signal reliably. Once again, biology has us beat as the homing pigeon can sense notes 400 times lower.

As for finding certain locations, the low frequency spectrum of the pigeon might not help it much as it does not update rapidly enough to create a stable signal. Instead, it is believed pigeons make use of the Doppler effect,[84] where the frequency of a sound is altered depending on the direction the sound source is taking. The most commonly used example might be a bystander listening to an approaching fire truck: as it comes nearer, the frequency increases due to the vehicle moving together with the soundwaves, making it sound like the siren's alarm is speeding up. When the fire truck has passed by and is leaving, it moves away from the frequencies picked up by the pedestrian, and the siren's frequency appears slower. In a similar way, when the homing pigeon moves towards an object, the sound it emits will increase, and conversely decrease if the object instead is moving away.

The anatomy of the pigeons' inner ear is also adapted for detecting infrasound. The nerve fibres have a high spontaneous discharge rate, meaning that at rest there is an even signalling to the brain saying that nothing of notice is going on.[86] As the spontaneous activity is of such a high frequency itself, any alterations to it will be detected quickly, allowing the pigeon to detect audible changes quicker than most other animals. The hair cells of the inner ear are also more similar to normal human skin hair than our cilia, grouping several hair cells into one single nerve fibre. This grouping means that a cell may be activated for a wider range of frequencies than the 1:1 ratio in humans and will react stronger to tones that are of very similar frequencies, with the side-effect being that this makes them less able to detect differences in pitch.

There is certainly no want of animals worthy of making an appearance in this chapter. Human hearing might hold up rather well but compared to animals who use it as a complement to vision, we are often left lacking. It does however make you think about the way we perceive our own senses. A bat probably does not consider its echolocation a way of seeing through hearing, and we might instead ponder how humans developed a way of hearing through sight.

Noses in the Animal Kingdom

Smell is one of the oldest senses made available to Earthly organisms. As such, it has had ample time to evolve over the millennia and the way it is enabled across various species offers an incredible example of the array of solutions for how we detect things in our surroundings, especially considering how many different eyes we have despite vision being a relatively new sensory concept. As it turns out, olfaction has remained remarkably similar throughout the progression of life, perhaps because of its early manifestation as a sense establishing it as a robust physiological platform. Smell is essentially the capacity to detect and correlate chemical substances through dedicated receptors. While this is largely conserved throughout evolution, the main variation in how this is achieved between animals seems to lie in how these substances are collected so that they may be detected or how the receptors are organised. Most people are aware of how dogs and many more animals possess far better senses of smell than us humans and you only have to look at the twitching nose of a mouse or the powerful trunk of an elephant to realise that there is much to wonder at about the animal sense of smell.

A Helping Nose

While we might think of man's best friend's keen sense of smell as invaluable to humans, dogs' ability to smell out our settlements might well have been one of the greatest threats to mankind throughout our evolution. Our human senses are generally pretty average. A pack of wild dogs circling in on an unsuspecting tribe would easily detect any guarding sentinel before they could see or hear each other. Lacking smell as biological radar, humans would have to overcome this threat by other means, but in a way we never did. When hiking in the mountains, we are still forced to seal food packs so as to not attract one of the world's best smellers: the bear. We have learned how to adapt, but while we can use binoculars and microphones to see and hear over distances unfathomable to other animals, our control over smell remains basic at best.

As illustrated in a previous chapter with dogs being trained to detect diseases such as Parkinson's disease, we have been successful in harnessing canines' incredible olfactory potential in different ways. But while dogs might be the best known olfactory aids, they are far from the only ones, and their competitors might surprise you.

The Wasp Hound odour detector is another example of humans training animals to aid in the detection of dangerous scents.[137] This contraption essentially consists of a cage with wasps. Being less friendly than dogs, the wasps have been carefully conditioned to react to certain substances by swarming. One study saw that, after having made some dubious associations with sugar-water, the wasps proved highly reliable in detecting cocaine, even being able to detect the powder despite being masked by tea leaves. Perhaps the next step will be to recruit police-wasps in law enforcement? Imagine the sweaty brow of a bank robber just after hearing that threatening echo from outside: "Release the wasps!" Still, the dog remains our most trusted aide in humanity's attempt to identify things by smell, which I consider to be very fortunate as I much prefer these four-legged friends over domesticated wasps.

Dogs do not only have more olfactory receptors than us humans, a whole 300 million compared to our pesky 5 million to be exact, but their anatomical structure is specifically designed to promote the sense of smell.[138] As a matter of fact, canines have two separate passages for air, as you can see in Figure 8. While humans have to rely on the air we breathe into our lungs to also pass by our nose, dogs have separate systems for breathing and smelling. A separate nasal cavity stores some air while allowing the dog to continue breathing normally, increasing the chances of airborne molecules to stick to a smell receptor. The wet nose is also covered in a layer of mucus which, much like our own sticky fluid inside our nose, traps smell particles. Similar to the way our hearing helps us identify the direction of a sound, dogs' nostrils work independently of one another. As a result, they can identify which nostril is closer to the source of a scent, providing them the tool for creating mental smell-maps that help in tracking. This canine olfactory prowess is further reflected in their olfactory cortex being 40 times bigger than ours.[138]

Dogs' true strength, together with that of many mammals and reptiles, lies in their *vomeronasal organ*.[139] This is the organ which picks up and interprets pheromones. As previously mentioned, humans lack a functional vomeronasal organ but retain a vestigial reminder of what-could-have-been in the form of an abandoned canal near the nasal septum.[140] In more adept animals, the vomero-nasal organ, also called Jacobson's organ after its Danish discoverer, is situated in the soft tissue of the nasal wall.[139] Much like the olfactory region, this organ picks up smell molecules. However, pheromone smell particles differ from the ones we can detect. Being essentially airborne or waterborne hormones, these particles will produce an instinctive response. This could be to hunker down in anticipation of oncoming predators, prepare the body for a deadly hunt, or anticipate a visit from a potential mating partner.[141]

The way these pheromones are processed is similar to how we integrate smell. Sensory receptors picking up pheromone molecules line the inside of a small cave within the canine nasal septum. Vomeronasal glands fill this space with fluid which allows the hormones to be dissolved, similar to the function of our own nasal fluids.[139] The grotto is also lined with blood vessels which, through their pumping, push nearby pheromones into this submerged space. Much like olfactory receptors send their signal to the olfactory bulb, the pheromone receptors relay their message to the accessory olfactory bulb. While the human olfactory nerve has the honour of being Cranial Nerve One, its canine counterpart responsible for hormone detection is instead given the more esoteric name of Cranial Nerve Zero, or the terminal nerve.[142] From the accessory olfactory bulb, the pathway is rather similar to that of smell with information being relayed to the amygdala and hypothalamus, both responsible for producing reflexive responses aimed at increasing survival. Far from being hidden, this response can be quite visible. For instance, when faced with an interesting pheromone, many animals will produce a *flehmen* response, borrowed from the German word for curling the upper lip.[139] When you see a horse do precisely that, it is because the animal tries to facilitate the transportation of pheromones into its vomeronasal organ. Many other animals do the same, ranging from cats to donkeys and even rhinoceri.[143] As different pheromones cause different responses, a skilled horse-whisperer would quite likely be able to deduce what type of invisible communication is going on under the farmer's unsuspecting nose.

Different Uses for Smell

Some might say that this detection of pheromones in animals could be considered a sixth sense, being distinctly separate from normal olfaction. The border between sensory systems is even more blurred in smaller animals such as the noble mouse. This animal has been shown to have a rather fascinating method of integrating scents. Sound and smell merge into joint units within the mouse brain, creating what scientists wittily have named *smound*.[144] In one study, researchers measured cell activity within the mouse olfactory cortex, where signals about odours traditionally should end up. When exposing them to combinations of smells and sounds, it was found that 65% of cells lit up to smells and 19% to sounds. Why sound is relayed to the smell-centre is unclear but the results showed that 29% of certain nerve cells had additive properties, meaning that they fired stronger when both smells and sounds were presented.

Focusing on this a little more, the fact that sounds alone can trigger signals in the olfactory cortex is in itself an interesting story. Does this mean that mice smell sounds? Do certain vibrations of air smell different from others? While strange to us, such phenomenona are far from impossible. There are several cases of human savants describing similar experiences. Savants experiences this form of *synaesthesia*, a combination of senses, for instance by seeing mathematical figures in different shapes, colours and intensities.[145] This has allowed individuals with such abilities superhuman mathematical skills as well as an amazing affinity for languages and writing. Despite their small stature, it would seem that mice, being born with the synaesthesia of hearing and smell, are similarly remarkable. It would seem that as the oldest sense, smell is a rather robust, and apart from the deviations in detecting pheromones and allowing synaesthesia, the neural principles of olfaction are rather well-conserved in vertebrates. The most notable difference is instead the relative importance this sense plays on our behaviours.

Dogs and mice certainly make good use of their great sense of smell, but naturally there are other animals equally dependent on olfaction. Even if you have never encountered one, I am sure you are familiar with the malodour associated with skunks. Using most animals' affinity for smells against them, the skunk pretty much lacks natural predators. The only regular enemy it has is the great horned owl,[116] maybe because skunks are yet to develop ground-to-air stink missiles. The truth is that skunks are very hesitant about using their weapon of choice. Their two anal glands produce the toxic spray, containing chemical compounds called *thiols* that smell excruciatingly of sulphur.[117] They have to be precise in propelling this out though, because after six uses the ammunition has run out, needing over a week to refill, leaving the skunk defenceless and without ammunition.[118]

Thankfully though, its reputation precedes it, making other animals give it a wide berth due to its olfactory deterrent. Interestingly, skunks generally do not spray other skunks, except for warding off other males during mating session.[119] This may be because the skunk does not seem to have any defence against its own weapon and should a skunk accidentally spray itself, their experience will be worse than that of a human smelling it, owing to their more sensitive sense of smell.

The Olfactory Elites

An animal that could easily avoid skunk territory, should it choose to, is the bear. Regardless of their subspecies, be it Grizzly or Kodiak, these apex predators sport the most reliable sense of smell in the animal kingdom. Their olfactory system does not seem to differ in any major way from other mammals, yet it allows these bruins to construct impressive internal maps of their surroundings.[120] Part of the reason why bears stand out in scientific studies is that their behaviour is easier to study in their natural habitats. A notable study investigated the olfactory capacity of brown bears by fully sedating them and moving them far out of their usual habitat.[120] One group was relocated 65–120 km away from their homes, while the other was instead displaced 120–270 km. The theory was that the latter group would have no visual cues telling them where to go. The results revealed an impressive impact of the bear's sense of smell on orientation: 65% of the less-displaced group moved homeward, compared to 74% in the long-distance group. Smell, it would seem, played an even larger part than vision.

Bears may not be the fastest predator, but their sense of smell helps keep them at the top of the food chain. Rather than being a hardwired instinct, the use of smell for guiding orientation seems to be a learned behaviour as cubs lacked the same smell-based proclivities to orientation.[120] Even in the icy climes of the Arctic, the bear nose reigns supreme. Here, its importance is even more prominent, which might not be surprising considering the lack of any visual means of orientation amongst the ice sheets near the North Pole. A grey seal lying on the ice may be hidden from view, but rarely from smell as the polar bear can sense it from 32 km away.[121] Even when underwater, the seal will have to watch out, as the air bubbles popping up through slits in the ice can be smelled from up to 800 metres by their predators.

There is one apex predator so adept at using smell for its high-octane hunting that it has even lent its name to a mathematical principle. The Shark Smell Optimisation model is based on its namesake's impressive ability to use smell efficiently when moving through the waters during a hunt.[146] In human engineering, its application allows for maintaining a balance of energy in electrical power systems, which in its biological counterpart translates into the most developed olfactory bulb ever described.[147] The size of a shark's brain is amply matched by its olfactory cortex, with two thirds of the brain dedicated to smell. Compare this to the one fifth of our human brains which is dedicated to vision. We use our eyes for mostly everything we do, yet it is not even comparable to the importance smell plays for the shark. It's importance is further illustrated by sharks quite literally not being able to stop smelling, as water continuously flows through their nostrils so their olfactory detection is not even affected by a breathing pattern.[147] A shark can afford no delays in its smell-based hunting, and their olfactory sensitivity is high enough that even a drop of blood entering the ocean over a kilometre away will be picked up and processed, sending these prehistoric hunters on their way.

Noses Big and Small

As we draw to the close of this chapter, I thought we would focus on two very different smellers. While one certainly looks the part, you might be surprised that the other even has a sense of smell. They were both featured in the previous chapter thanks to their unique methods of hearing and it would seem their sensory abilities continue to impress. Much like their approach towards hearing, the elephant and the moth approach smell in very different ways.

With its long trunk, the elephant certainly looks the part of a world-champion smeller. Its genetic material supports this notion as the African Elephant with its 1,948 smell-identifying genes has the highest number of odour-oriented genetic segments ever discovered.[124] This translates to a great number of olfactory receptors lining the inside of the elephant's trunk and it is commonly said that *an elephant never forgets*, which is certainly true for smells as elephants can maintain an impressive long-term memory of different smells. A study in Kenya revealed that they can even distinguish between ethnicities of different local humans.[125] On a more practical level, an elephant's sense of smell helps them identify pools of water in the otherwise desiccated African landscape during dry seasons.[126]

Moving away from the world's largest land-dwelling animal, we enter the insect world. Here, the moth, despite its humble size and non-existent nose, offers an even more impressive capacity for odours. What they lack for in frame, these survivalists more than make up for in terms of smell. It has been shown that the male silkmoth can detect an isolated pheromone molecule from a female of its own species over 9.6 km away.[123] Combined with their amazing sense of hearing, the moth would certainly make a formidable villain in any science-fiction novel. Rather than possessing a nose however, the insectoid counterpart in moths consists of their antennae always being on the lookout for biological residual matter.[129]

While the moth can thank its hearing for being able to swoop through the air avoiding bats, most other aspects of their biology are dictated by their sense of smell. A male moth will find its mate by her scent and the female moth will in turn choose where to lay her eggs based on the fragrance of plants.[130] Since they are so reliant on smell, moth behaviour is very predictable, which is exploited in the agricultural industry. Moth larvae inflict significant damage to crops, so farmers often implement measures including scent manipulation to confuse potential infestations.

Smelling Through Someone Else's Nose

The moth also helps us answer a question none of us knew we wanted the answer to: what would happen if we transplanted noses with a different person? Would we have a new favourite smell? Would we experience the smells just like the original owner, or does our own brain overrule it all? A study from 2016, was able to successfully transplant the antenna of one moth to another without destroying its capacity to detect scents.[148] The answer to the question of whether the nose or the brain decides the nature of a smell came as a surprise: it is neither. Different subspecies react differently to certain smells, but after transplanting antennae between them, the reaction of the moths suddenly became unpredictable, leading to the conclusion that a brain is primed for its particular nose. A change in input leads to a changed reaction, but as the central command has not been prepared for this, the outcome is something completely new.

If such a relationship holds true for our other senses, it would appear as if our sensory organs harmonise only with our own brains, creating a distinct individual pattern that is changed if any part of us were replaced. Perhaps this answers the age-old and somewhat philosophical question you sometimes come across: could it be that we see colours slightly differently, but nevertheless call them by the same name as we learn their association during childhood? If you and I were to change eyes, perhaps the colour I perceived as orange you would call red, or vice versa. Drawing such parallels based on a moth's sense of smell might be difficult, but it does raise the question on just how subject to interpretation our senses truly are.

When you experience a smell as different from someone else, it is most likely because the way you and your nose are wired, and the associations you make have been moulded by your past. While we may expect a more uniform response between individuals with visual or auditory stimuli, these other means of surveying the world are also subject to calibration through nurture. More so than other senses however, smell remains such a personal experience which we understand the world cannot share. Smell reminds us that we are all unique, and our biological prejudice renders the world unique to us.

Taste in the Animal Kingdom

Despite being physiologically rather straightforward compared to senses like vision and touch, taste does a pretty good job of illustrating just how relative life might seem. It makes you wonder how other life forms experience their source of nutrients: Does the oak tree enjoy the taste of sunlight? It is difficult to think of light having that type of quality, but then again we eat spicy food simply because we essentially enjoy the taste of pain. Compared to most other organisms, humans have a pretty well-balanced conception of taste. The reason for this is quite obvious: we are omnivores and eat a little bit of everything. Consequently, we need to be able to taste a little bit of everything. As we shall see, the capacity for detecting different tastes is reflected by an animal's culinary needs.

Tasting with Your Genes

Had we, like cats, been incapable of tasting sweetness, we would probably never have perfected the art of sugar refinery to the degree of having 39% of all adults in the world be overweight.[149] We have arguably developed beyond what would be considered our evolutionary boundaries, as our struggle for survival has in times of increasing food supply instead led to a growing obesity epidemic with associated mortality. Perhaps some insights and solutions can be found in the feline DNA. Your neighbourhood cat might not be able to enjoy sugar but this is nothing more than a cruel trick played on him by its genetic code. A major gene coding for our ability to taste all things sweet is the T1R2 sequence.[64] While all felines have this, it is completely turned off within their cells. It therefore seems pretty likely that early cats were able to indeed taste sweet. After a couple of millennia, the gene went dark, becoming what is referred to as a pseudogene. Still, we should remind ourselves that not all things we perceive as sweet are limited to activating only one set of taste buds, and that olfaction also plays a crucial part in a food's flavour profile. Cats may be able to detect some of this through smell, but will always fall short of the full sensation due to their different taste receptors.

So why would a meat eater even have this gene in the first place? The answer is of course that at some point in the cat's evolution track they must have benefited from consuming carbohydrates, but as time would tell their path through the food-chain lay somewhere else. As gene therapy continues to progress, might it not be conceivable that an efficient way of dealing with obesity may be to simply downregulate our ability to taste sweetness? After a few generations, we would not even miss our candy bars and pastries. After all, we cannot miss that which we have no perception of.

Birds have taken a heavier hit than cats in Nature's crusade on the senses; their T1R2 sequence has been completely removed.[150] One explanation for this may be that they used to eat fruit but some shifted from picking at apples to digging for termites. This could have been advantageous for many reasons, and eventually a genetic predisposition towards this behaviour would build up.

One biological discrepancy, however, is that some birds actually do seek out sweet substances. Hummingbirds, for example, are notoriously fond of sugar water, though snobbish enough to be completely uninterested in artificial sweeteners.[150] Despite this, there is no sign of that all-important gene coding for the perception of sweetness. So how do these animals hunt for sugar when they should be incapable of detecting it? The answer lies in the protein sequences lining the sweet-tasting T1R2 gene, with these being logically named T1R1 and T1R3. From a human perspective, we have these particular combinations of amino acids to thank for our perception of umami. Initially, like all birds, hummingbirds were unimpressed by sugar. Somewhere along the line, they likely discovered that nectar was a pretty good source of energy and those who could seek it out were evidently more successful. This change was likely brought on by a mutation in their umami-tasting genes, making them less sensitive to umami but instead capable of detecting sweet carbohydrates.[150]

As a point of reference, humans are guided by about 4,000 taste buds, turning chemical substances into electrical impulses just like those of all other animals. By comparison, the worst taster in the world was long thought to have much fewer gustatory receptors. After some digging, scientists were able to detect 24, then 70, until finally settling at the still meagre 240–360 taste buds.[151] Unlike the bulb-shaped taste buds of most animals, these avian creatures have egg-shaped little feelers, which is ironic considering that the worst achiever in this particular field is none other than the chicken. Like other birds, chickens also distinctively lack the ability to taste sweetness because of their DNA. Figure 9 offers an overview of some notable tasters as indicated by their number of taste receptors.

In addition to their unseemly low number of taste buds, only 2% of them are located on the chicken tongue, with the rest having set

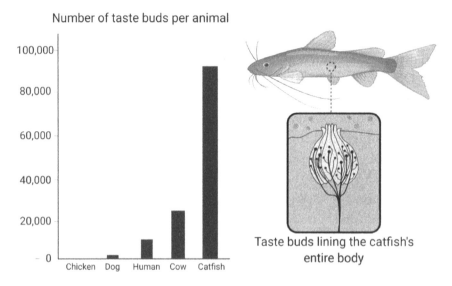

Figure 9. Capacity for taste. The capacity for taste across the animal kingdom. The number of taste buds reflects an animal's affinity for detecting taste, and the chart to the left offers some context for how well humans perform compared to other animals. The catfish, illustrated to the right, has a great affinity for tasting thanks to the high number of taste buds spread across its body. Figures created using Biorender.

up camp in the rest of the oral cavity.[151] By the time a chicken can taste whatever they have put in their mouth, it is therefore a bit late to spit it out. So what is the point of taste if it does not help you separate the healthy from the harmful in whatever you're ingesting? It seems chickens simply use it as a daredevil's version of memory: they eat something, they get sick, hopefully they survive and next time they probably will not eat the same thing. These birds therefore have to rely on other senses to guide them through the menu, particularly vision.

Evolution seemingly played around with birds such as the hummingbird before deciding that the original model was pretty good after all, first getting rid of their sweet-tasting gene and then creating a new version of it. As mentioned before, this sequence will not respond to artificial products like aspartame, probably because of its curved evolutionary pathway. After all, aspartame was adopted by humans in order to replace sugar as a low-calorie sweetener but from an evolutionary perspective, it has no business pretending to be sweet. Tasting aspartame as sweet would most likely be detrimental as it would essentially be seeking out a drug which has no caloric value. This accident of evolution may be due to aspartame's ability to bind to the protein-compound that makes up sweet taste molecules. This example beautifully highlights the fickleness of sense perception. We do not enjoy aspartame because it is actually sweet, rather it is sweet because we have created it to be perceived as such. The inherent qualities of biological life are rarely those that we impose upon them through wilful manipulation.

The capacity for detecting different tastes serves as a reflection of an animal's culinary needs. For instance, giant pandas are known to mostly only munch on bamboo. This might not only sound pretty bland, it is also generally a bad idea as bamboo contains next to no nutritional value, although it is plentiful. Pandas therefore have very little time to do much of anything other than eating, and as a result they have little use for tasting non-bamboo foods. Much like cats turned off the gene for sweetness throughout their evolution, pandas turned off the sequence coding for umami perception.[140] The perception of an animal's environment is therefore dictated by a species' genetic coding, which in turn is influenced by the environment the species inhabits.

Surviving Without Taste

Some animals stand out even more by nature of their gustatory disinterest. Much like chickens, sea lions have a very short time span to detect taste molecules as they swallow their fish whole. Serving no obvious purpose, the DNA coding for the perception of sweet, umami, and even bitter foods have been turned into pseudogenes and rendered useless in sea lions.[152] So how do these animals avoid poisonous foods? Where taste falls short, the other senses step in. Sea lions are excellent at visually identifying stuff they put into their mouths. Even more impressively, they can tell what animal they are hunting based on the vibrations carried through the water as picked up by their whiskers.[153] While chickens taste by colour, sea lions by contrast taste through touch. There is no merging of sensory information; a chicken's optic nerve is not intertwined with the nerves innervating the tongue. It is all a matter of association or evolutionary conditioning. Much like we enjoy flavours based on a mix of smell, touch and taste, other animals too judge their intake by multisensory data.

Dolphins and whales hunt in much the same way as sea lions and as a result, they have also undergone a significant loss of taste receptors.[154] Not to be outdone by their sea lion neighbour, this did not stop after losing only three taste receptors. Due to their tendency to eat krill and other types of fish whole, cetaceans have lost all sense of taste apart from salt. Conceptually, this might seem strange considering the salty waters they inhabit; imagine if we humans could taste nitrogen, which makes up about 80% of air. Presumably whales and dolphins adapted different reference points to what could be considered salty and they do not notice the smell or taste of their own habitats. So what makes salt so special that it survived the Great Extinction Event of taste buds in cetaceans? Arguably, it must be for the same reason as for all other animals: salt is imperative for retaining homeostasis. Even if you have evolved past such frivolous things as tasting food, you will still need an internal reminder to even out your fluid levels. We have touched on this before, but remember that without it our cells would cease to function. Salt, therefore, is just as important as water.

Interestingly, there are some evolutionary idiosyncrasies within the cetacean family of mammals. For example, it is quite feasible that whales lost their taste receptors when they stopped chewing their food, which in turn is why they do not have teeth but rather a long row of bristles called baleen.[154] Dolphins by contrast still have razor sharp teeth, the function of which is not for chewing but as weaponry. Dolphins and whales have however done quite well for themselves despite their bland feeding habits, mostly thanks to the astuteness of their other senses. However, much like cats, cetaceans were not always bereft of other tastes. Having evolved from an animal called the raoellids, a land-dwelling mammal that looked much like a possum,[144] they were exposed to a much wider variety of foods than in their current habitats. They changed diets from berries to fish sometime around 50 million years ago, as they went further and further into the Indian Ocean.[144] This is pretty inspiring: Nature creating a whale out of a possum is probably the best illustration of the power of evolution.

So are there any downsides to having essentially no sense of taste? For most of existence, not much it would seem. However, being the planet's most invasive species, we have significantly altered our environment, with consequences often being detrimental for the survival of other species. Humans have spilled so much waste into the oceans that whole islands of trash are reported to be floating around. Lacking a conventional sense of taste, species such as cetaceans are none the wiser when they move through oil spills or consume fish that have been poisoned through industrial waste. If nothing changes soon, we can only wait and see if cetaceans can reactivate their silenced taste buds before it is too late.

Taste by Seeing and Seeing by Taste

Humans are guided by about 4,000 taste buds, turning chemical substances into electrical impulses just like those of all other animals. The worst taster in the world was long thought to have much fewer gustatory receptors, in fact none. After some digging however, scientists were able to detect 24, then 70, until finally settling at the still meagre 240–360 taste buds.[145] Unlike the bulb-shaped buds of most animals, these avian beasts have egg-shaped little feelers, which is ironic considering that the worst achiever in the auditory field is none other than the chicken. Like other birds, they also distinctively lack the ability to taste sweetness because of their DNA.

In addition to their low number of taste buds, only 2% of them are located on their tongue, with the rest setting up camp in the rest of the oral cavity.[145] By the time a chicken can taste whatever they have put in their mouth, it is therefore often too late to spit it out. This begs the question of what the purpose of tasting is for chickens if it does not help them separate the healthy from the harmful. It seems chickens simply use it as a daredevil's version of memory: they eat something, potentially get sick, hopefully survive and next time they encounter the same thing, the unpleasant memory associated with it will discourage them from eating it. Much like sea lions then, they rely on other senses such as vision to guide them through the menu. They even have visual preferences as well, so if you want to play it safe when offering a chicken a snack, make it red.

Thus far, we have only seen how other animals come up lacking in having fewer taste buds compared to humans. This might be a nice change as it is pretty rare that we come up on top in the sensory game. Sadly though, we cannot claim the crown this time either. Unlike its feline namesake, this fish has retained all its basic tastes, as well as possessing a whopping 175,000 taste buds.[155] The catfish resides in muddy waters where vision does not do it much good. So in this case, the taste buds have taken the role of eyes, allowing these bearded vertebrates to identify surrounding objects by their taste. Most of the catfish's taste receptors are interestingly not in its mouth but rather spread over its entire body, so in all fairness you could argue that this makes the catfish one giant tongue. The boundaries of taste and smell might seem a bit blurry at this stage, they are after all just versions of detecting a chemical signature. Generally, one can think of the catfish tasting particles which physically make contact with a receptor, while smells are detected by its long whiskers when in contact with waterborne chemicals.

Touch in the Animal Kingdom

Touch is a rather universal concept. The neural physiology behind it is, for the most part, not as divided between species as that of hearing for instance, which sports echolocators that use it almost as if it is a whole other sense. This uniformity is rather telling of just how basic our need for touch is; even if the sense has gone through convergent evolution, it is difficult to tell this from just looking at the way animals behave today. For instance, humans and octopi have very different eyes but our tactile limbs carry the same type of receptors, allowing us much the same in the way of tactile sensation thanks to their similar neurophysiology. There are however some differences to animals' neural implementation of touch to warrant some special mentions.

Touch by Hair

You may have noticed that there are a great many animals who sport impressive sets of whiskers. These tactile hairs, vibrissae, have a rich neural network hugging their roots and a follicle-sinus complex where six distinct groups of neural recorders stand ready by the hair's root to collect all vibrations it picks up.[77] The use of whiskers is a fundamental fixture on the mammalian face, with exception of primates. This does of course not mean that whiskers are any less important. We would do well to remember that from the perspective of natural history, we are just starting out as a species and we may never have risen over the level of footnote had it not been for us prioritising brain size over most other things. As we have seen before, most animals express very different priorities when it comes to what constitutes the most important sense of perception. For touch, it is particularly animals that live in darkness that sport the most impressive sensory apparatus.

The naked mole rat is a good contender for illustrating their importance. All body hair has fallen off this underground rodent,

rendering it something akin to a mythical creature. Its dark esoteri-
cism does not stop there, considering that it is seemingly impervious
to both cancer and the onslaught of age; furthermore, as it can survive on extremely low supplies of oxygen, the naked mole rat might
appear as almost immortal.[156] For us, the supernatural tendencies of
the mole rat are however secondary, since this otherwise naked creature still retains its whiskers. While you might not call it fur, the
animal cannot really be considered naked as it also has small tactile
hairs covering its entire body; if the catfish is a swimming tongue,
the mole rat might be likened to a giant fingertip. Despite its tactile
sensitivity, it completely lacks nociceptors in its skin which makes it
essentially impervious to pain. There are many hairy creatures that
deserve mentioning in this chapter, and some more approachable
than the mole rat. While cats and mice sport more familiar whiskers,
the general principles of touch-by-hair remains largely the same
across the animal world, and you can find the general principles laid
out in Figure 10.

Some animals are hyper-sensitive to touch. Much like the naked
mole rat, arachnids also have hairs which allow them their superhuman somatosensory prowess. The hairs covering their entire body,
called trichobothria, sense airborne vibrations and even electrical
charges.[157] The latter might seem like a separate sense in itself, but
one may argue we are also capable of detecting electric fields in
much the same way, as illustrated by rubbing a balloon against your
arm which will make the hairs rise in response to static electricity.
These numerous little hair cells allow spiders to gather all-important
information about the surrounding world just by standing still.
Vibrations in the air, or whatever surface the spider is in contact
with, create a rustle along the thin projections making up the spider's fur. Carried on through the stem much like the hair cells
allowing us the sense of hearing, the movement reaches the touch
receptors. A flexible membrane hugging the hair follicles enhances
the motion to better allow the receptors to fire up a charge in the
associated nerve.

**The Touch-Me-Not
(Mimosa Pudica)**

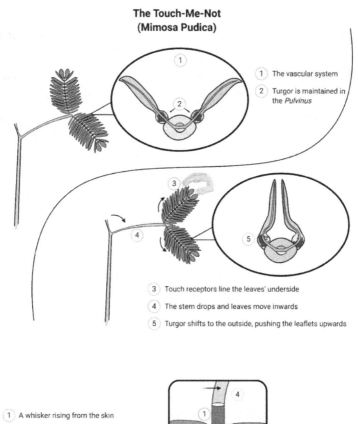

1 The vascular system

2 Turgor is maintained in the *Pulvinus*

3 Touch receptors line the leaves' underside

4 The stem drops and leaves move inwards

5 Turgor shifts to the outside, pushing the leaflets upwards

1 A whisker rising from the skin

2 Superficial whisker nerve

3 Deep whisker nerve

4 A movement of the whisker causes tension further down the stem. The two nerves can then pinpoint the hair's exact location

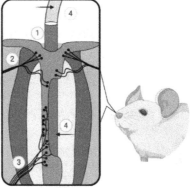

Figure 10. Different forms of touch. The top image shows touch in the plant kingdom. The Touch-Me-Not responds to tactile stimulation by altering the water turgor in its cells. The shift results in the leaves and stems moving closer together. The bottom image illustrates the whiskers of a mouse, with hair cells representing one of the most common forms of touch in the animal kingdom. The schematic shows how a whisker detects touch and how tension in the embedded portion of the hair will engage different sets of nerves. Figures created using Biorender.

Considering how the spider uses its hair cells, you might label this version of touch as hearing. It is after all hair-borne vibrations that allow soundwaves to be translated into electrical impulses for the arachnid brain to comprehend. I find this to be a beautiful example of how our view of the senses is moulded by our expectations of them. A spider might never have suggested that hearing and touch were different senses. Had we asked it about the concept, it might instead have gotten terribly perplexed. Imagine our confusion if we are visited by an alien race asking us which type of solar flare we prefer to enlighten our third eye. We might be able to detect solar flares but we have no biological concept of how it affects us. In the end, it makes you wonder what other senses are out there that we humans are missing out on.

To close this segment on hair, let's lightly touch on some water-dwelling mammals. As water relays vibrations much more efficiently than air, the vibrissal system of hairs is key for aquatic orienteering. Just from visual inspection, this is not so hard to believe since the walrus arguably sports the most majestic set of whiskers found in the animal kingdom. While it has often been joked that Salvador Dali could paint with his impressive facial hair, the walruses actually do use theirs as wispy rakes, combing the muddy seabed for clams.[77] Another blubbery sea-dweller, the magnificent manatee, has whiskers all over its body much like the naked mole rat. Bending to the water's fluid motion allows the manatee to position its massive body in the best position to brave the currents.[77]

Unlikely but Touchy Organisms

We risk being very hair-heavy in this chapter, which just goes to show how important this method of perceiving touch is. However, let us leave fauna behind and step over into the world of flora. Consider a leafed plant nestled close to the ground. A pink round flower protrudes by its elongated leaves and when you look closer, you notice how the finger-like leaves are actually made of smaller ones, sticking out from the central stem like miniature versions of the larger group. As you touch the thin green tendrils, they contract inwards, reshaping the group of tiny leaves into a thin spear as if to escape your touch. This is the Mimosa Pudica, or the "Touch-Me-Not". While there are several plants that respond to touch with what is called "rapid plant movement", the Touch-Me-Not is the most famous, having been migrated around the world by people intrigued by its somatosensory perception.

So how does a plant devoid of muscles, joints or any type of neural system that could explain this basic animalistic sense detect touch? The cause behind it is still somewhat vexing, with the current theory briefly outlined in Figure 10. Plants generally do not move and by contracting its leaves, the Mimosa Pudica uses up a lot of energy, so the reason behind it must be important to favour it over the traditional floral defence-mechanism of standing still. It is generally believed that the small movement might be enough to scare away some herbivores, and even humans. Quite possibly, the movement might also serve to shake off any unwelcome houseguests trying to eat its leaves.

Unsurprisingly, the mechanism through which this floral ambulation works is quite different from any human counterpart. Leafy cell walls are usually kept in shape with the help of water in what is called turgor pressure.[158] When the plant touches an unknown object, the cell walls will release a set of chemicals. Having the balance within a cell disturbed, the water will suddenly flow out of some cells and into others to correct the water levels. As water levels drop, some cells of the leaf will collapse. Neighbouring cells will also feel that something has changed, repeating the chemical procedure and soon the entire group of leaves have changed their shape.

As you may have realised, none of the animals above employ the same pressure-sensing system as us humans. Another creature which has rejected hair-mediated somatosensation for specialised cells embedded in the skin is the *Hirudo medicinalis,* or medicinal leech which has for hundreds of years been used in bloodletting for medicinal purposes.[159] Despite having a tiny neural network, only employing some 10,000 neurons, we may have more in common with this invertebrate in how we perceive touch than one might think.[160] As a brief introduction, the medicinal leech is a blood-sucking parasite which for hundreds of years have been used in bloodletting. Historically this has been used to treat a number of diseases, often making the patient sicker than when he first sought help, but the practice does continue to this day for some disorders where there is a pathological overproduction of blood. I mention this because it highlights the purpose of this leeches physiology: seeking out prey from which it may extract its sanguine cuisine. Much like in humans, leeches have specific cells dedicated to touch, pressure, and pain. This network allows the leech the impressive range of movements it needs to swim, climb, and manoeuvre around its prey for maximal suction. Having a segmented body, its soma-tosensory network enables each segment to respond with an immediate and precisely calculated response, while also informing the remainder of the body what is going on. Despite the hundreds of millions of years that separate us from the annelid phylum to which leeches and worms belong, the comparison between human and leeches can be quite humbling, as our central nervous systems appear remarkably similar with its ladder-like layout of sensory neu-rons reaching out from a central column. In reviewing evolutionary crossroads, David Ferrier sums it up quite concisely by stating that "Man is indeed but a rather fancy worm".[161] Once again, we see Mother Nature's power and creativity reflected in evolutionary convergence.

The world of touch is clearly a widespread one, ranging from the highly specialised neural network of mammals to the simpler mechanical actions of spongy hair cells and floral leaves. As the old-est sense, it has guided all aspects of life on this planet since the first strands of RNA bound together in the stormy oceans billions of years ago. It has been argued to be the sole human sense, the foun-dation on which the other four rest. Even Socrates himself viewed taste as a form of touch and I personally think that a point can be made for our sense of hearing being considered a variation of the somatosensory perception of vibrations. You could of course argue the converse, that the various aspects of human touch are too different to be considered part of the same sense. Personally, I feel the current classification works best. Having to describe the touch of a loved one in terms of vibrations and thermal fluctuations makes it lose some of its intrinsic meaning. Touch is perhaps the one language that all living things on this planet can agree on, so perhaps we can cut Aristotle some slack and keep this monster of a sense as it is. Things are, after all, complex enough as they are.

Chapter 9

The Superhuman Senses of the Animal Kingdom

Products of Time and Space

Let us for a moment leave our human-centric view of senses behind and seek inspiration in what was never meant to be for our fragile forms. Have you ever looked up into the skies and imagined what life would have been like should nature have gifted us wings? I would wager that you have. Imaginatively allowing ourselves the power of flight, or any other superhuman ability, is such a universal experience we humans go through growing up. Flying is probably the most obvious ability that biologically must remain forever unattainable to us humans, as birds cruelly flaunt their avian superiority in plain sight. Sadly, as we shall see throughout this chapter, it is far from the only ability refused to us. Superhuman abilities have always been a point of fascination throughout history. Essentially all religious and cultural mythologies reveal as much, from the incredible strength of Hercules to the shapeshifting Loki. Today, superheroes are making a comeback to such an extent that several storylines have had to be reset to manage the menagerie of heroes and heroines. Quite often, their attributes are borrowed from the animal world, and with good reason.

The world is filled with strange and mysterious phenomena, which even we as humans struggle to make sense of despite our advanced technology. It is equally evident that these phenomena are not the same for all of Earth's inhabitants. Much like we would struggle to imagine a world without smells, the humble bumblebee might think us crazy not being able to appreciate a good electric field. Imagine then what other strange forces lurk out there in the vast darkness of space. We are a product of our environment, and as humankind attempts to extend the limits of our species, it is crucial that our sensory capabilities must advance. Technology has come a long way and amazing progress is being made by studying the animal kingdom. While we strive to adapt our computers to pick up on seismic activity or fluctuations in the Earth's magnetic field, nature has spent millennia perfecting similar skills in our animal neighbours. So let us now take a trip through the sea, fields and skies. While we may share this planet, Earth is not the same to all of us.

Exploiting the Polar Opposites

Let us start off with a closer look at those aviators with their coveted ability to roam the heavens. Many birds migrate vast distances every year with the record-holding Arctic tern flying a distance of more than twice that of Earth's circumference over a year.[162] With such vast distances to cover, how do they know where they are going? It took us humans thousands of years to master the art of navigating the seas, and only recently have our technological devices advanced to the point where they might actually match the biological compass of migratory birds. Let us here introduce our first superhuman sense: magnetoreception. Just like gravity, Earth's magnetic field has been there from the very start of biological life. It takes some travelling for it to shift to any noticeable degree however, so it would make sense that only animals that perform distant journeys would develop any affinity for it. Several animals have done exactly that, but for now let us stick to birds. As it stands, there are two biological components to this sensory ability. In order to successfully orientate themselves along the magnetic field, birds need to both identify the direction of the said field, as well as deducing its strength.

As with all magnetic fields, our planet's magnetic field is created by electrical currents. They are produced by streams of liquid iron in the planet's core caused by the immense heat trying to make it to the surface.[163] This circular current produces two magnetic poles that extend their influence through a magnetic field stretching into space where it collides with particles emitted from the sun; the resulting auroras can be seen with the naked eye in the far north or south where the magnetic field is closer to Earth's surface. As a result, we have strong magnetic activity near the poles, and a magnetic needle floating on water will align itself to this north-south axis. This is exactly how compasses work, and why they point north, though ironically the geographical North Pole is the magnetic South Pole and vice versa, as the geographic and magnetic pole locations differ due to the fluid and changing nature of our planet's molten core. In addition to the intensity of this axis, the magnetic field also has variations in inclination and declination, referring to the angle made by a compass at any given location. As we have seen previously, understanding basic physics is essential for understanding life, so armed with this knowledge we can now move on and investigate how birds take advantage of Earth's magnetic field.

It is generally believed that a bird's ability to detect the direction of a magnetic field resides in the eyes, while the intensity is calculated by features in the beak.[164] Both these concepts are illustrated below in Figure 11. This is not to say that birds can see magnetic inclination, or smell its strength. Rather, they use the same sensory organs we humans associate with vision and olfaction in ways that our human biology cannot accommodate; perhaps beaks are to birds for magnetoreception what our mouths are for taste. Either way, these two perceptions of the magnetic field are physiologically quite different and could almost be considered two distinct senses in themselves. As a means for explaining the directional component

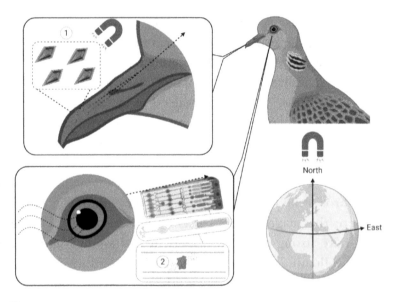

(1) Magnetite in the bird's beak adjusts according to the planet's magnetic field, signaling its direction to the brain

(2) The protein cryptochrome sits between the photoreceptive discs at the back of the retina, hosting the radical-pairs compass when interacting with light

Figure 11.   A superhuman sense. Several animals can detect the Earth's magnetic field, using it as a compass for navigating vast distances. This process can be illustrated in the pigeon, where the two major methods for magnetoreception are outlined. Created using Biorender.

of sensing the magnetic field, we have the radical pair model.[164] As the eye receives photons in the form of sunlight, there are certain proteins in the retina that form radicals, meaning that they have an odd number of electrons in their molecular structure. According to the radical pair model, this is not an isolated event; instead, the protein is matched with a twin through quantum entanglement, meaning that the pairs are perfectly synchronised in the way they behave. You can think of one of them as a very defiant twin, doing the exact opposite of its sibling so that when their activities are added up, you get an absolutely neutral zero. Just as a magnetic field is created by the flow of electrical forces, the direction of these paired radicals produces a magnetic moment which can be detected by birds' retinal proteins, called cryptochromes. In short, the direction of the radical pairs produced by a photon hitting this protein will point in a specific direction depending on its location relative to Earth's magnetic field. While not a perfect analogy, imagine a spinning top being kept going by the forces of its own rotation. Now imagine giving it a little nudge with your finger. If you were quick enough, it probably tilted a bit without falling over. If you were to film and analyse the change in its rotation and tilt, you could probably tell from where the push had come. I imagine the beak-compass to be somewhat similar, although more complex and certainly more reliable. Still, it is still unclear how the resulting signal makes it to the brain, and we do not really know if magnetoreception is integrated in the same area as vision or if it has been dedicated a distinct brain area of its own. However, a study has shown that if a bird's eyes are exposed to different magnetic directions, the bird will prefer to follow the signal detected in the right one.[164] As the optic nerve crosses sides when entering the brain, we can at least be pretty sure the signal is processed in the left hemisphere.

To truly make use of the magnetic field, birds also have to deduce its magnitude, as it fluctuates depending on your geographical location. This involves the second aspect of magnetoreception which involves something as unique as internal biological magnets called magnetite.[164] Depending on the way they are clustered, magnetite have different affinities in aligning with external magnetic forces. While their exact location has been disputed, modern imaging techniques sensitive to electric and magnetic activity have revealed that in pigeons, these particles reside in birds' upper beak. Anesthetising this area has also been proven to produce some very disoriented birds, seemingly cutting them off from their magnetic compass. One of the most conclusive pieces of evidence for magnetite's role in magnetic navigation was seen in an experiment which saw birds exposed to brief pulses of magnetic radiation.[164] While humans cannot detect this, when the birds were let loose, it quickly became apparent that they were not quite as unaffected. Migratory directions were seemingly random over the next days, until the birds' internal compasses had realigned themselves and it took ten days for them to re-adopt their normal flight pattern. Should there ever be a remake of Hitchcock's horror-classic Birds set in modern times, perhaps neutralising the avian threat with an electromagnetic pulse (EMP) may be both an unexpected and efficient solution.

The neurophysiological basis of how magnetite's influence is relayed to the brain is somewhat better understood than the radical pair hypothesis. Measuring the trigeminal nerve, the same nerve that relays sensory information of pain and makes spicy foods seem painful in humans, has revealed increased activity when birds were exposed to changing magnetic fields.[164] As the name indicates, the trigeminal nerve has three branches, and it is specifically the ophthalmic protrusion slithering around the eyes that is responsible for the surge in activity. Damage to the trigeminal nerve has also been linked to birds no longer being able to adjust their course during migration, providing further evidence that this nerve is vital for carrying magnetic information. As for higher brain areas involved, there are several candidates. The vestibular nucleus, responsible for balance, has for example been shown to have increased activity during magnetic exposure. So have the thalamus, hippocampus and hyperpallium, the latter being the avian version of what we mammals call a cortex. Interestingly, this activity has been traced back to signals transmitted through the lagena. This structure, which humans do not possess, is a third otolith organ next to the saccule and utricle.[165] Its function has been much debated in amphibians and lizards, though it seems to be involved in self-orientation considering its position in the inner ear. Only in birds has distinct evidence of its purpose been found, with researchers finding that lesioning the lagena causes brain activity associated with magnetoreception to be significantly reduced.[164]

While the role of the vestibular system in magnetoreception certainly is interesting, it does not seem to be necessary when detecting magnetic fields. As stated at the beginning of this chapter, birds are not the sole proprietors of this sense. Today's infrastructure, supporting our many cities with vast power grids supported by cables passing through forests, streams and fields, carries powerful electrical currents and subsequently distorts the magnetic field in their vicinity. Both cows and deer have been shown to align themselves in particular patterns when grazing underneath these power lines.[166] Considering the relatively short time humans have been around, this sense most certainly hails from times when Earth belonged not to humans, but was free for its inhabitants to roam its vast biomes. Clearly the ability is useful, and Earth's magnetic field will very rarely let you down if you need to find your way over great distances. Human industries would quite possible never have developed had it not been for the compass, but there are those who claim that our ability to detect magnetic fields goes back even further, hidden in our very DNA.

Let me start off this segment on human magnetoreception with imploring you to take it with a healthy pinch of salt. Interesting studies are certainly being made in the field but are generally inconclusive. With that being said, some evidence points towards us humans possessing the same magnetoceptive protein that birds have: cryptochrome.[167] Based on this, one could infer that the radical pair model should hold true also for us, and that our retinas could carry electrical charges outlining magnetic directions the same way birds can. We should however remember that humans lack the lagena otolith, which seems to play a key role for our avian friends. Nevertheless, a recent study even claims that humans are capable of utilising this phenomenon in pointing out directions, albeit only East and North, and only if said humans are starving and male.[168] In this study, subjects were seated in a metal cage through which a magnetic field could be imposed, giving researchers control over magnetic direction as well as making sure participants could not be aided by knowing their geographical direction. Initial attempts revealed no human magnetoreception, with people pointing out directions at random. The researchers argued that perhaps this directional ability is latent, and only useful during threat of starvation, as it could aid in localising previously encountered food sources. Participants were consequently told not to eat for 24 hours before returning to the lab. Once again, the results were disappointing. Researchers now started priming their subjects by offering them snacks in certain magnetic directions, hoping to trigger a primordial response that would help them stave off this self-inflicted fasting. After realigning the magnetic field, it was found that these primed subjects had a significantly higher success rate in identifying north and east, but only if they were men. The published article concludes that the ability to have magnetoceptive powers awoken when blood glucose levels reach a certain level disappears if no blue light is present in the room, which is certainly interesting as the whole concept of magnetoreception is built around a protein in the eye.

Guided by Electricity and Pressure

We do not have to leave the skies quite yet, as our next animal has been known to fly across oceans despite lacking wings. Previously, scientists believed this was through hitching a ride on wind currents. Instead, it was found that these fussy land-dwellers make use of Earth's electric fields, which is produced by the different charges between Earth's surface and the atmosphere and can be readily illustrated through thunderstorms.[169] The species we are talking about is spiders, and their fuzzy bodies come to life with static electricity to such an extent that it can actually propel them through the skies.[170] Spiders' affinity for electricity was proven by studying them walking around in a confined chamber. As an electric field was introduced to the space, the hairy inhabitants started tip-toeing around, and some even took off in a brave defiance of gravity.[170] Arachnids are consequently the second arthropod species to our knowledge that can sense and make use of the global atmospheric electrical circuit. Bees, equally hairy, make up the second half of these sensory adepts. Bees build up a positive charge in their hair cells when flying, and as plants are negatively charged, this helps them locate their much coveted nectar through literally being attracted to the flower.[171]

Darwin himself had brought up static electricity as a possible explanation for spiders' strange prowess in aviation.[170] The theory remained dormant for several hundred years without gaining much traction until recent studies brought it more to the fore. In an artificial environment, consisting of a cardboard strip between two oppositely charged metal plates, spiders have been shown to express the same pre-launch behaviour when the field was activated as Darwin himself had noticed in the wild: they stand on their toes, stick their posterior into the air, and then off they go. Experimenters have even been able to affect the altitude at which the arachnids levitated based on the strength of the electric field. We previously talked about the sensitive nature of a spider's hair cells, the trichobothria. The purpose of these hair cells exceeds that of simple touch, as they flutter in the electrostatic winds undetectable by our human senses.

Without leaving the field of electroreception, let us take a quick dive into the oceans and visit a species that has roamed the deep sea for a very long time, longer in fact than there have been trees on land.[172] Sharks are exceptionally sensitive to electric fields.[173] Their sensory prowess extends so far as to detect the charge from a standard battery through electrodes set up 16,000 km from each other. The purpose of such a fine sense for electrical fields is the same as all other functions of the shark: hunting. Fish generate small electrical signals when moving about, something they themselves take advantage of through using it as a means of communication. Unfortunately for the fish however, there are few obstacles disguising their electrical footprint, and like the Big Brother of an oceanic surveillance state, sharks are hot on their tracks. Shark electroreception is made possible by tiny little tubes filled with superconductive gel.[173] Scattered across their heads and hidden underneath a thick layer of skin, they provide their owners with ample information on all aquatic movements. These electroreceptors, named ampullae of

Lorenzini after their 17th century discoverer, consist of a bulb at the bottom of each gel-filled tube. As the electrical current moves through the conductive gel, it reaches a membrane in the ampulla, which in turn activates the nerve that plugs in from underneath. In this case, the electrical current is simply carried onward to the brain almost like a continuous stream. Compared to the other senses, this seems pretty efficient, considering that the likes of smell and hearing have had to find other ways, usually through mechanically poking hairs, to translate energy into electrical impulses.

Without leaving the oceans, let us look into the next sense: pressure. Sure, we humans may also feel pressure through touch, or more abstractly when faced with starting an essay due tomorrow. We are nothing but amateurs though, as our aquatic ancestors had a whole sensory system dedicated to detecting pressure changes of the sea, which in a terrestrial context would translate into a human being able to discern the height of a hill just by atmospheric pressure. Water-dwelling vertebrates have something called a lateral line system.[174] Mechanoreceptors, not unlike those responsible for our sense of touch, line the fish, amphibian, lamprey and other aquatic vertebrates from head to figurative toe. These neuromasts, as they are called, form interlocked rows along their entire bodies, and quite often the receptors are hidden away underneath the skin in long jelly-filled trenches. It is this setup that ended up being used for electroreception in sharks.

Polarised Light and Biological Sunglasses

Sensing pressure might feel very similar to sensing regular touch. I would however argue that the mechanism is unique enough to merit calling it a distinct sensory system. The same thing can be said for other aspects of aquamarine life. We have already discussed octopus vision and their uniquely advantageous eyes, but there is even more to it than meets its cephalopodan eye, bringing us swiftly to the next superhuman sense.

So, moving away from fish, imagine holding a gentle octopus. To dig deeper into the mysteries of octopus vision, we need to talk about the nature of light. You might have heard about polarised light, which is often associated with a specific type of eyewear — polarised sunglasses. These are increasingly popular, particularly amongst fishermen as it allows them to better see below the waves. As you might have guessed, this means that the light under normal viewing conditions is unpolarised. Now, light is a rather complex matter and so we will have to simplify it immensely for the sake of space. We have previously talked about how particles can have a spin to them, such as the model explaining magnetoreception in birds. Light consists of electromagnetic waves carrying the photons that eventually hit our retina, and these photons have a spin which can be either to the right or to the left.[175] In unpolarised light, this is generally balanced, while polarised light waves have a spin in only one direction. The reason why we cannot spy things underneath the surface of a lake is because water acts as a natural polariser. Water has a tendency to appear rather flat at rest, so the reflected polarised light is only filtered in a horizontal direction. Polarised sunglasses add a vertical polarisation filter, which brings the photon back to being more clearly discernible for our human vision as their spin is once again balanced.

Vertebrate retinas are nicely arranged in a semi-random pattern with respect to the photosensitive rods and cones. Consequently, we do not discriminate the direction of any incoming light since the retina covers all possible inclinations. By contrast, the photosensitive proteins of octopi and other cephalopods have a much more deliberate orientation, whereby they have a directional preference for the spinning photons; essentially, their retinas work as a biological pair of Ray-Bans.[176] Similar to perception of pressure in fish resembling perception of touch, this perception of polarised light could in a way be considered as a sub-specialisation of vision. Still, I would argue that the ability to discern photon spin is enough to warrant light-polarisation a spot on the non-human sensory tapestry.

Our understanding for how this perception of photon spin is possible however is still being developed. While not precisely proven, it seems logical that octopi make use of this ability when navigating the seas, particularly when hunting or communicating with other cephalopods.[176] Researchers have also found that cuttlefish, another cephalopod, sport fancy markings across their body only visible in polarised light. Picture yourself in a classroom taking a test where you and your friend are sitting across the room from each other. Imagine the possibilities if could you project the answers to a difficult question on your forehead without the teacher being able to pick up on it. Clearly the sensory ability to detect polarised light carries with it important advantages in a world where stealth is your best chance of survival.

Moisture, Air Pressure and Natural Meteorologists

Thanks to their aquatic environment, our cephalopod friends have no need for our next sensory system. Water is essential for all organic life, so the ability to find it is crucial for survival. We have already talked about elephants being able to sniff out water over large distances, but that is just another aspect of smell. Beyond this, some animals have evolved a distinct sensory system for detecting humidity and moisture called hygroreception.[177] Unlike the cephalopods' ability to view polarised light, there are specific receptors associated with this ability. In that way, there is certainly no doubt that finding humidity constitutes a well-rounded sensory system, albeit one denied to the human species.

As many other elusive senses, the method by which hygroreception works is shrouded in somewhat of a mystery. Three models have been put forward: mechanical receptors detecting proportional humidity, evaporation detectors that respond to a decrease in internal saturation, and lastly through psychometers which rely on comparing drops in temperature between dry and wet hair cells.[177] It might well be that various animals implement different strategies in their hunt for moisture. A study in cockroaches for example revealed that the latter method involving temperature was the most likely candidate. Using UV light, researchers were able to detect the activity of those cells responsible for detecting moisture changes.[177] By controlling the environment in a very precise manner, it was found that the cells expressed the most reliable reaction to changes in temperature. It would therefore seem logical that the hygroreceptors are sensitive to the loss of heat when water evaporates from the surface of the cockroach hair cells. So how would that work? If you have ever spilled pure alcohol on your hands, you might be familiar with the rapid cooling effect it has. This is because as the alcohol evaporates, it absorbs heat from your skin during its transformation from liquid to gas. The same concept is true for this hygroreceptive model: when the air is dry and the temperature high, more energy is needed for the water to evaporate, and vice-versa. Genetic studies in the fly have also shown that several receptor types code for both humidity and temperature, indicating a close relationship between the two.[178] There are however receptors sensitive to only humidity, and not temperature, which indicates that there is also a direct way of surveying the surroundings for humidity, though again this may well differ between animals. Consequently, the mechanical model of detecting absolute humidity levels in the air has also gained much traction.

How, then, is air humidity able to inflict mechanical stress upon a biological system? In another study on the cockroach, researchers focused more intently on the insects' antennae. Together with the stick insect, the cockroach was able to reliably detect changes in air humidity as well as pressure.[179] This was tested by monitoring cells embedded in their protruding feelers, which proved rather adaptable. As the moisture in the air increased, the relatively dry cells absorbed some of the water through osmosis. This is the process through which most of the fluids in our body transport in-and-out of the bloodstream: when we lose salts through sweating, there is an increased concentration of salts left in the solution, so to balance this issue, our cells release some of their fluids into the general bloodstream until both blood and cells have a roughly equal water-salt ratio. As a result, the body can rely on a steady stream that reaches all organs equally. For the cockroach, this means that the dry cells in its antennae will absorb water from the humid air. Quenched in the airborne nectar, each cell starts to swell, pushing on its neighbouring structures.[179] This is where the mechanical aspect of this model comes into play: as the cells simply push on the nerve located next to it, the cockroach's brain receives information on the air's humidity in direct relation to the amount of pressure inflicted by its water-absorbing cells. Simply put, higher humidity means bigger cells, which means a bigger push on the nerve. Its ability to sense pressure arguably works in a similarly mechanical manner: an increased pressure means that each cell is compressed under the change in atmospheric weight, which in turn means that the neighbouring nerve cell has more room to expand, causing a change in its signalling as the forces on the nerve adjust accordingly. This latter phenomenon has a generally weaker impact on the cockroach's cells compared to humidity however.

One could argue that if the ability to detect pressure through the compression of one's cells is possible, then surely humans also possess this sense. From a completely mechanical perspective, this is arguably true: if the surrounding pressure were to increase drastically, say by having a boulder fall on us, then we would certainly feel that through our gift of touch. Any atmospheric disturbance would leave us unaffected however due to our skin cells simply not being sensitive enough to allow us these powers. Clearly our sense of touch has not evolved to deal with these aspects of our environment, even though we clearly are affected by both humidity and pressure, although not to the same extent as other animals. This biological purpose is key when discussing senses. For a trait to truly be considered a sense, it must have evolved for that specific purpose; fish sense pressure through their dedicated lateral line, while humans might accidentally guess at a drop in air pressure as a by-product from our other senses. One quite popular example of the latter is the wise old grandmother whose knees start to ache whenever there is rain incoming. While the reason behind this is somewhat unclear, it is argued that lower air pressure, mixed with cold temperatures and some gentle rain, can have enough effect on the tissue within an arthritic knee to cause its unruly nerve endings to increase their signalling.[180] Grandmas are however not designed predominantly to foresee the weather. Their mystical powers should instead serve as a reminder for the rest of us as to how sensitive their poor joints have become.

Chasing Flames

As with all things, there are of course grey zones for what can be considered a proper sense, and we have already seen plenty of examples. One additional such challenge is presented by some subspecies of the colourful Jewel beetle. Gifted with a structured exoskeleton that would make any designer cry with delight, their beautiful iridescent patterns are produced by a complex geometry that allows only certain wavelengths of light to bounce back from any point of its body. Heavily contrasted to their environment, this insect relies on destruction to survive. Almost poetically reflected by the gleam in its blazing carapace, the Jewel beetle needs to seek out fire.[181]

Unkempt forests tend to bog down the natural forces that reside within, but much like all other biological systems, these too need to make room for future generations. Similar to how the world is prophesized to rise again after the great sundering of Ragnarök in Norse mythology, fires can transform old forest into hotbeds of new possibilities. In the fertile ashes, new woodland will rise, inviting new migrants from the neighbouring glens. The Jewel beetle cannot afford to let such opportunities pass; the larvae grubs need the rich soil left after burning trees in order to get the nutrients they need.[181] Consequently, the Jewel beetle lineage would be but a streaky smear on the windshield of history had they not developed an innate affinity for finding fire. Note however: pyroreception is not a well-established concept, but I would suggest such an attribute could be awarded this rather exceptional insect. To detect fire, the Jewel beetle harnesses the combined powers of touch and smell. More precisely, they make use of thermoception, and not even smell might prove the correct term; beetles use receptors on their antennae to pick up on particles carried by the smoke.[181] In this regard, one could certainly argue that they correlate to our human noses, but as these feelers hold far more utility, direct comparisons between our two species are not exactly straightforward.

So what is the underlying mechanism for pyroreception? From a chemoreceptive point of view, the antennae of the Jewel beetle can certainly smell a burning woodland, which happens simply through detecting airborne particles associated with fire.[181] Their olfactory ability allows the Jewel beetle to detect burning wood from over 80 km away, even allowing them to differentiate between tree types. Regarding their thermoception, it would seem that it is less important for finding fires than orienting around them.[182] Despite their love for flames, the Jewel beetle is not immune to their destructive nature. In order to escape a sudden fiery death, our adventurous insect is armed with a system of detecting infrared radiation. Flanking both sides of its abdomen, this organ allows its owner to detect fires the same way you might have seen police pick up heat signatures using specialised cameras. You may recall the snake's pit organ from the vision chapter, which also functions much in the same manner. Researchers long thought that these organs were to detect fires, but a recent study however saw beetles trying to escape from infrared radiation shone on them by scientists.[182] Considering that covering their infrared-detecting organs with aluminium foil deprived the insect its pyroreceptive capabilities, it seems likely that the Jewel beetles' affinity for sensing fires is indeed backed up by their biology.

In the end, it is difficult to say whether the Jewel beetles' ability to detect fire can be attributed to a sensory system or not. Smell and infrared detection can certainly pull their own weight, and from a physiological point of view, these are the two systems in charge for finding fires. From a strictly behavioural perspective however, I would argue that the concept is far from a done deal. A sense aims to gather information about the surrounding world, and it would seem that detecting fires is the end-goal of this particular mashup of biological mechanisms. In the end, there is something very specific about the way the Jewel beetle is designed around this one phenomenon. Perhaps one could say that pyroreception is their main sense, and we only think of it as smell due to our human-centric standpoint. How many smells must a nose be able to discern before we go from calling it a designated particle-detector to an actual olfactory organ? I will bravely shy away from answering that question. Nevertheless, the Jewel beetle reminds us that the borders between the senses are not written in stone.

The Dorsal Light Reflex and Using the Sun for Balance

Our next item on the list is similarly located in the sensory No-Man's-Land. In the previous chapter, we talked about balance as a sixth human sense, especially focusing on the vestibular organ. Of course, balance is the complex child prodigy of vestibular, visual and somatosensory information, but in terms of having a specified organ associated with this all-important ability, we must give credit to our inner ear as the primary sensory organ. As we have already discussed, the method through which we humans, and most vertebrates, achieve this is by assessing the inertia of our inner ear fluids. Considering that all animals must be able to maintain some sort of balance control, it might come as somewhat of a surprise that not all of them possess this ability. Still, insects fly the right way up so there can be no denying their sense of balance despite lacking the appropriate organ. Instead, they make use of a different system for determining which way is up. Many fish also possess this non-vestibular system in combination with our coveted inertia-sensors[183]. Still, the need for balance is older than the vestibular system, and so it is not surprising that some species further down the phylogenetic tree might possess several solutions to one problem. Instead of using only inertia as a means of detecting balance, they make use of another physically reliable phenomenon: incoming sunlight.

Despite the different inclinations light has as the sun soars across the sky, it is generally true that the sky will be lighter than the ground. Consider a fish, resting by the sandbank of a shallow stream. Its bulging eyes give it an almost comical expression, pointing outward to either side of its elongated head. As the sunshine breaks the surface of the water, the prismatic light comes into contact with the cells of the fish retina. In its stationary position, the illumination of the two eyes is equal in strength. Now, consider instead if our fish had been slightly tilted. Now more light will reach the side that is facing the surface, while the other eye is shaded by the head. In an automated response to this misalignment of the incoming light, the fish tilts its body until the sun is safely above its scaly head. The same concept holds true for invertebrates such as squids and many insects, revealing an ancient physiological response imprinted throughout the biological spectrum. Some evidence even suggests that humans retain some sort of dorsal light reflex, although only appearing in cases when children have grown up with occluded vision in one eye.[184] With one retina shrouded in darkness, the weak light-mediated balance response signals to the brain of a perceived head tilt for such a long time that the eyes actually end up slightly rotated. In the end, this produces a slightly tilted image in the retina that the brain has to readjust.

This opens a similar question to that brought on by the Jewel beetle, although inversed. In the beetle's case, we had two senses aiming to produce one specific goal, finding fire. Here we instead see how one sense, balance, is accomplished by two different physiological mechanisms. On the other hand, one can claim that the vestibular organ is the actual sense and balance is just the desired by-product. Considering the importance of vision and somatosensory information on postural control, this might be a fairer interpretation. If that is the case, then we need to think of a better way of referring to this sense. Calling it "vestibular signaling" is rather clunky. Personally, I would argue that inertia fits the bill quite well, as it describes the organ's primary function. The dorsal light reflex could then be considered a sense in its own merit rather than a version of balance. It is true that eyes are involved, but in a strict sense the reflex does not involve vision in any way as it does not rely on producing an image; instead, it simply involves comparing the lighting conditions in two separate photoreceptive areas. Therefore, while the eyes make up the organ, the responsible neural mechanism is different enough to detach it from the concept of vision altogether. I will however accept that this is essentially a matter of perspective which is open to debate.

Feeding off Radioactivity

Let us move on to something a bit more exotic. Humans have had a colossal impact on essentially all aspects of this planet. It is however rare even for us to threaten its very existence. One such exception might have been the Chernobyl disaster of 1986. Had things gone even slightly worse, we might have been living in a reality with a completely inhabitable Eastern Europe. As it stands, 100,000 square km were contaminated.[185] Even today, an exclusion zone of 30 km remains around the main reactor. While humans are kept out for the most part, other animals are less keen on following rules. Examples like this highlight the potential usefulness of being able to sense dangerous levels of radiation. Radioactivity is not a human invention after all. Every second, Earth is bombarded with background radiation from outer space. Clearly we have evolved to withstand this status quo, and our atmosphere protects us from most of it anyway. Nevertheless, some minerals are more radioactive than others, with uranium being perhaps the best-known example of

Earth's deadly rocks. Avoiding these traps would arguably be a pretty decent trait, especially if you live close to the ground.

Let us travel back into the Chernobyl Exclusion Zone. One might expect that an animal sensing radioactivity would stay clear of such a nuclear hotspot, but in 1991 a radiosensitive organism was found to inhabit not only the general area, but the very epicentre of the nuclear power plant itself.[186] These radiotrophic fungi thrive in the high-energy atmosphere surrounding the nuclear fallout site. Rather than using photosynthesis, the act of translating solar energy as a form of nourishment, three different species of fungi make use of gamma radiation to grow.[187] In a way, this might not be as strange as it first sounds; the sun is after all a giant nuclear reactor which is only harmless to us because of our evolutionary distance to it, and if we one day move to Mars, we will have to find a way to avoid the nuclear energy produced by our closest star. What stands out is the speed at which the fungi evolved to their new surroundings, just about 35 years. Evolution is dependent on previous generations passing on genetic variations, and the life cycle of a fungus is a lot shorter than that of humans, which partly explains this phenomenon. Another key factor is that there need to be genetic deviations that can be passed on in the first place, which can happen through recombining the parental genetic material or through random mutations. As you may know, the Chernobyl disaster caused, and continues to cause, severe types of cancers through its destruction of animal DNA. While this has caused a drop in the life-expectancy of most animals in the Exclusion Zone, it also sped up the mutation rate needed for our fungal evolution. So while most plant life perished, the radiothropic fungi emerged as the apex life form of their new habitat.

The chemical mechanisms behind this high energy diet remain largely unknown. Photosynthesis traditionally works through having chlorophyll use sunlight to break up water molecules. Radiosynthesis instead seems to involve an increased production of melanin. As a natural pigment, melanin dictates such things as skin or eye colour, but for the Chernobyl fungi, melanin might be the key ingredient in their survival. Melanin namely acts as a support for a series of metabolic events, and ionising radiation has been found to enhance this capacity; a specific radiotrophic fungus even tripled its growth when exposed to this destructive discharge.[187] For now, this process remains a bit of a mystery, but its implications might soon become all-the-more important. Space travel is for example full of dangerous radiation. If we could harness this fungal superpower, then perhaps growing food would be made much easier in space as the otherwise deadly beams from the sun instead could improve the growth of plants, or maybe we could just eat the fungus itself.[188] Much like plants pick up toxic particles in the air on earth, a fungus feeding off radiation would also act as a radiation sponge, providing cheap and important protection from the unseen dangers of space. From an evolutionary perspective, this ability to feed off radiation might also explain early life on Earth.[189] Ancient unicellular life lacked the protective dome of Earth's ozone layer. It is generally believed that the unfiltered radiation from the sun forced life to the depths of the oceans, but perhaps fungi just like these were the first to take hold on solid ground. Through using melanin in their metabolic processes, they could have set the stage for future life, as the fungi produce oxygen and other products we so desperately crave. Having set the stage, other animals would eventually follow the radiothropic organisms and give rise to brand new ecosystems.

Contrary to this fungal philosophy, eating radiation is generally considered a bad idea. So what about other radiosensitive species? One could surely argue that animals that risk getting too close to Earth's toxic crust would benefit from some sort of internal warning system. As life would have it, rats seem to have developed precisely that. More specifically, they seem to be able to "smell" radioactive material.[190] Though to be fair, it is difficult to say if it really is experienced as a "smell", but from a human perspective it is probably our best point of comparison. Researchers have known this for quite some time, and it was revealed in a study where some rats were exposed to radiation and presented with saccharin water, while a control group was allowed to enjoy their sweet treat in peace.[190] The hypothesis was that if rats were able to identify the dangerous radiation, then they would show aversion to saccharine, as long as they belonged to the group having been conditioned to it. Results revealed precisely that — a single exposure to radiation during the experiment was enough to make rats shun the saccharine water like poison. Smell was shown to be the associated sensory system as rats that had undergone complete lesioning of the olfactory bulbs, the region of the rat brain integrating scents, failed to learn this important lesson.

Humans arguably have little use for detecting ionising radiation. During our evolution, any rock we would have accidentally stepped on would probably have been too small to affect us anyway, a luxury not afforded our smaller vertebrate cousins. However, even rats have smaller organisms surrounding them. So does that mean one's need for sensory systems is inversely proportional to an animal's size? One can see how bigger animals could be less dependent on sensory systems as their sheer size would deter any would-be aggressor. Naturally it is not this simple, and a species' sensory needs have to adapt to the neighbours they keep, with large prey often being subject to even larger predators as they compete in an evolutionary arms-race up the food chain. I bring this up because I am going to introduce a rather small guest at this point, not for its sensory prowess but rather for its complete and utter disrespect for the whole concept.

Beyond All Senses

While rats can sense radiation because it is harmful to them, this animal would be none-the wiser even if it had been airdropped directly into Chernobyl at its worst. Our next bear-like micro-animal would have no trouble traversing the deep sea before immediately visiting the peak of Mt. Everest. They live in the Arctic as well as in the mud of volcanoes. They can set up shop in either arid deserts or leafy rainforests. As far as we know, this animal has only touch as its means of interpreting the world as it is armed with hairy bristles along its head and body. I would argue that a lack of senses can be just as descriptive of an animal as the ones they command, and for that reason the tardigrade deserves a special mention at this point.

Tardigrades, also known as moss-piglets or water-bears, are the most resilient species of animals known to science.[191] In the greater scheme of things, one could argue that tardigrades are the true masters of this world, having been here for at least 530 million years according to fossil records, remaining unchanged despite living through several extinction events.[192] Considering that they have been shown to survive even the desolate vacuum of space, proven in 2007 by having a satellite deposit a group of tardigrades in orbit before collecting them again, they are seemingly near indestructible. We might be limiting human survivability through the irreparable damage to our global climate which threatens mass-extinction, but if the moss-piglets were to take over, they would undoubtedly happily scurry about the radioactive fungi we may end up leaving behind.

Water-bear is also a very apt name for the tardigrade considering their plump physique, as you can see in Figure 12. Being only half a millimetre long, they have four pairs of distinct, albeit stubby, legs ending in impressive claws.[193] Despite their size, they sport a brain with multiple lobes and a neural network to carry out its will.[194] As of their mysterious past, it has been theorised that they underwent a shrinking stage throughout evolution, with ancient caterpillars being their original form.[195] As previously stated, they make use of hair cells, cilia, in their tactile encounters, but seem to care little about any other sensory input. They do however possess eyes, probably inherited from their earlier caterpillar form, but these now reside within their brain, and many tardigrade species are completely indifferent to light.[196] I find that tardigrades, through their relatively sensory-deprived existence, reflect the fickle nature of

Tardigrades

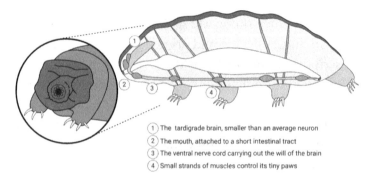

1. The tardigrade brain, smaller than an average neuron
2. The mouth, attached to a short intestinal tract
3. The ventral nerve cord carrying out the will of the brain
4. Small strands of muscles control its tiny paws

Figure 12. The tardigrade. These animals thrive in extreme environments, unperturbed by noxious sensory input. The tardigrade serves as an evolutionary illustration of the proverb *ignorance is bliss*. Created using Biorender.

perception: we interpret only that which to us is important for survival.

There is no such thing as a supreme sense. Perhaps having too good a sense might even be detrimental; if moss-piglets could feel temperature, then they would certainly not be as undeterred in their exploration of space. In 2019, an Israeli lunar module crashed on the moon's surface carrying, among other things, tardigrades.[197] If what we know about these tiny animals is true, then it seems likely that the moon now is inhabited by terrestrial life, making the tardigrade the first non-human to visit another planetary body, and the first to stay indefinitely. Organisms that thrive in such extreme climates are usually referred to as extremophiles.[198] Moss-piglets fail to qualify for this description however, because they have not evolved to make use of extreme conditions. More importantly from an ecological perspective is that these small insensitive piglets might well have set the stage for other species to inhabit new areas, which, like the radiothropic fungi, would classify them as a pioneer species.[199] As tardigrades eat bacteria, they do not really need much in order to thrive. These survivalists go where others cannot, eventually attracting bigger lifeforms to follow suit and join a freshly developing ecosystem.

You might ask why we are finishing this sensory expose on a species so disrespectful of the sensory world as the tardigrade. They highlight perhaps the most important point one can make about the methods through which we survey the universe. Senses are there to guide us through a highly specialised environment. They exist to guide us through the habitat in which our ancestors decided to set up shop, and more often than not, these endeavours fail to be successful; indeed, 99.9% of all species have gone extinct throughout Earth's history.[200] I hope this chapter has given some insight into what living in different ecosystems and climates across evolution might have been like with reference to the senses their inhabitants had to develop. We have seen how a creature with echolocation like the bat lives in a pretty dark place, while creatures sensing the Earth's magnetic field seem to be pretty good at navigating long distances. Humans, on the other hand, are somewhat of a generalist species, the general practitioners of Homo sapiens. We might not be the best at much, but it is our diversity that has made us strong, although seemingly inferior to the moss-piglet.

Chapter 10

The Future of the Senses

It is tempting to imagine that we humans are no longer at the whims of Mother Nature, that modern society has developed beyond the inconveniences of our ancestors and evolution is something you only read about in natural history. But nothing is static, and now more so than ever are we reminded of forces beyond our control. As I write this paragraph for instance, the Covid-19 pandemic has been escalated by a string of mutations of virus. This rapid evolution serves as a reminder that life is ever-changing, and we are only passengers on this great journey.

If we truly are in a continuum of evolution, what does that mean for our senses? We have seen where they came from, but where are they heading? A common misconception about evolution is that the most desirable trait will always win out; in reality, the only factor that matters is how likely you are to pass a certain trait to an offspring. For that you need to produce offspring, who in turn have several children carrying the mutation. To use viruses as an analogy, the most successful virus is not the one that causes the most damage, but one which can live within its host and replicate the longest. If anything it would be disadvantageous to have the host expire, since that means less time available for the virus to produce new copes of itself. Humans are rarely reliant on our sensory prowess to produce offspring these days. Still, if we take a closer look at trends in human behaviour, certain trends emerge.

If we are capable of impairing our sensory capacities through loud sounds or poor reading conditions, what can we do to improve them? Will technological advancement and further industrialisation lead to an inadvertent drop in biological function for future generations, or can technology be used to improve our perception of the world even in the long run? What will the role of artificial intelligence be in all of this, as our abilities to create bionic systems and machine-brain-interfacing are exponentially improved? This chapter will take a brief look at some of these questions. Perhaps with some inspiration from the past and some imagination, we can try to make sense of what may lie beyond.

The Future of Vision

One thing you may have noticed when comparing the modern day world to life in old videos or photographs is that a lot more people are now wearing glasses. These spectacles are hardly new inventions, the Italian monk Salvino D'Armate created the first eyeglasses in 1285[201] and Benjamin Franklin is credited with having produced the first bifocals.[202] Aids such as these have become more readily available over time, both thanks to reduced production costs and increased availability of opticians, but part of the reason also lies in changes in our human anatomy.

Eyes are getting bigger, that much seems to be true. As you may remember from the chapter on vision, myopia is caused by the eye being too long for the lens to correct the incoming light. The resulting near-sightedness is what causes the increased need for eyeglasses we are currently seeing. This change has happened over a very short time-span, and in the US, the number of bespectacled individuals has increased from 25% in 1971 to 40% today.[203] In many Asian regions, this number has reached 80–90%.[204]

So what does this mean for the future of vision, and what is the underlying cause? Myopia is generally inherited, but we are also moulded by our visual surroundings as we grow from children to teenagers. The eye goes through two major growth-spurts, one during infancy and then again during puberty.[205] It is believed that our modern lifestyle is causing this growth to exceed that of previous generations. Increased use of screens and tablets, more time indoors, and more intense study periods all seem to contribute towards this change. It is sometimes said that development is evolution on turbo, and much like evolution attempts to promote a certain physiological feature, our eyes mature to fit the needs of their owners; if the visual world is never further away than a computer screen, then that is the distance for which our eyes will specialise. Focusing on objects near you, such as reading a book, will put an extra strain on the eyes. In order for the lens to accommodate, the ciliary bodies contract, and like most muscles these can get fatigued. Elongating the eye relieves the muscles of some of this work, and as a result we become better-suited for near-work, but struggle to see things in the distance.

There are several genetic factors that contribute towards myopia, which in turn may contribute to the condition only in the presence of environmental factors like the ones described above.[206] It seems likely, therefore, that the worsening of human vision stems mostly from our modern lifestyles. The good news is that we can try to alleviate this risk by keeping a watchful eye on the reading habits of our children and teenagers. This is important because the changed anatomy of the eye can not only lead to myopia but also several other disorders, such as glaucoma and an increased risk of the retina detaching.

All in all, the future of vision will likely involve continued deterioration so that requiring glasses becomes the norm, although it is unclear whether this will turn into an evolutionary divergence for the human eye. Nevertheless, we have the opportunity to move away from it by changing our habits. The future of human vision, for good or bad, remains in our hands.

The Future of Hearing

It is tempting to suggest that hearing may go the same way as vision. We are, after all, exposing ourselves to far louder sounds than our ears were intended to take in. For many people, having the loudest speaker system is a source of pride, and the consequent poorer hearing an acceptable casualty. At the same time, many workplaces are now far better equipped to protect us from loud sounds than say, the factories of the Victorian era. Luckily, there seems to be no clear trend of our human hearing changing on any grand scale. With that being said, there are reasons for listening to the scientists when it comes to the future of hearing.

In the previous chapter on hearing, we explored how early humans were adapted to different sound frequencies, largely thanks to the anatomy of their outer ears.[31] In this way, hearing may seem a likely candidate for future evolutionary modification, which naturally invites the question of whether there is any evidence for such changes currently taking place, and if future generations may struggle to hear our 21st century speech should they go through radio and TV archives.

As it stands, our ears seem to be staying in their current form for the foreseeable future; at least, we have no clear indication that our hearing is changing. Compared to vision, there is no obvious cultural or technological shift that has put hearing under a comparable stress. The ear is also less likely to adapt during maturation, so that even if your parents only spoke in a very specific frequency, it is unlikely that you would grow up to be unable to hear other people's speech.

There is however one important aspect of future hearing that scientists are becoming increasingly concerned about. Much like our changes in vision, this cause for concern is being brought on by changes in human behaviour. While it is hard to know if the phenomenon will affect our peripheral organs, it may well influence the central interpretation of sound, making us more sensitive to particular sounds. When discussing the future of hearing and sound, it is important we take a look to noise pollution.

Human civilisation is becoming increasingly urbanised and in many parts of the world, this is happening faster than the infrastructure can keep up with. It many ways noise pollution seems inevitable and was even documented as early as ancient Rome.[207] Cityscapes literally stack people on top of each other, and many of us will have been exposed to loud music well into the night. The dangerous levels of sound tend to come from the industries we share space with, such as construction work, electrical generators, sanitation, public transport, all of which make modern cities a never-ending cacophony of sounds.

So what does this mean for the future of hearing? The World Health Organisation has set the recommended upper decibel limit for residential areas to 50 dB.[208] Today, the average in many urban regions reaches 98 dB.[208] Prolonged exposure to sounds of this magnitude will damage our hearing, but it also increases the risk of cardiovascular diseases such as hypertension and coronary artery disease[209] as well as cognitive decline.[210] Many of us have likely felt the stress and hypersensitivity associated with exposure to continuous loud sounds, often becoming annoyed at things we usually have the mental reserve to ignore. Despite no physical changes having taken place, we become more sensitive to sound; not in the sense that we necessarily hear better, but we may be quicker to form reflexive emotional and physical responses to otherwise harmless sensory input.

Unlike our visual decline, we rarely implement personal gadgets to counteract the effects of sound. If anything, we try to drown out loud noises with even more sound by turning on our headphones. Cities need to take heed of the impact on noise pollution on health, and progress is indeed being made with silent roads, planned construction work, and sound barriers next to roads. The problem is that the cities most affected by noise pollution tend to be in developing regions of the world, placing an unfair weight on already burdened communities. The future of hearing will likely continue to remain tightly intertwined with how we build our living spaces, and there seems to be little we can do as individuals to mitigate these effects on our own sensory systems. What we can do, as always, is to be mindful of those we share our space with.

The Future of Smell

There can be no doubt that human exposure to smells has widened significantly across the millennia. Having spent most of our time in small bands as hunter-gatherers, we now live in communities and cities full of odours which our predecessors never could have imagined. It is also very clear that our sense of smell is moulded by our environment; recognising an old smell from childhood may evoke long-lost memories, and when we visit new places, one of the first things you may take notice of is how it smells. Simply put, our sense of smell is calibrated to our environment, and reminds us where we came from.

Humans originated in Africa and have since then made almost all regions of Earth home, regions that undoubtedly smell very different to each other. So in order to predict how smell may change, we can look to the past to see what evidence exists for this ancient sensory system having continued to evolve in modern humans.

Assessing the evolution of smell is challenging, given how we are moulded by our upbringing; for instance, a South-African will likely disagree with an Icelandic colleague on the nature of an odour simply due to their different points of reference. In order to test the evolution of smell, we need to go beyond individual perceptions. This brings us to the smell receptors themselves, allowing us some insights into how our DNA is steering our nasal sensors.

Androstenone is a steroid pheromone found in the saliva of wild boar. While humans lack the capacity to detect pheromones, we do possess receptors for this specific chemical compound.[211] At least, many do. A study carried out with participants from 43 different populations across the globe found that our sensitivity to this smell differs between people. As suggested by their DNA, participants were able to detect the scent of androstenone based on the genetic composition of its associated smell receptor (OR7D4).[211] Groups in Africa tended to be able to detect this scent, while those in the northern hemisphere did not. This north-south divide does appear to have been dependent on intermixing with other human species throughout evolution, such as the Neanderthals, since their genetic makeup suggests they were also able to detect the compound. It is perhaps a bit of a mystery why some would be able to smell andros-tenone and some not; perhaps it is just one of those random mutations that were passed on through chance rather than offering any concrete advantage for survival.

Despite smell being one of our more robust senses, it is there-fore clear that it has evolved even during the short existence of modern humans. Considering its intrinsic connection with taste, this also means that human flavour perception is prone to evolution. So what does this mean for the future of smell? It certainly suggests that our perceptions of odour may be changed by our genetic mate-rial. With that in mind, the fact that humans are becoming more and more connected, and isolated populations increasingly rare, makes it tempting to speculate that such changes between peoples will slowly disappear.

Regarding the future of smell, it seems likely it will carry on mutating as humanity continues its journey across the planet and beyond. Smell, perhaps more so than vision or hearing, is remarkably adaptive, and it seems unlikely that our future survivabil-ity will depend on it. Still, random but dominant mutations may appear in the future and may by chance be passed on through the generations. We know it has happened in the past, and it would be strange for it to never happen again. Ultimately, there is likely little we can do to influence this trajectory, and perhaps we have very little cause to attempt to do so. So let us tag along for the ride and allow our noses to take the lead.

The Future of Taste

It is quite tempting to group our sense of taste with that of smell when looking into the future of these senses. As already discussed, when we talk about taste we usually mean flavour, which is an amalgamation of several perceptions of a food or drink. Since our sense of smell is prone to changes, so is our interpretation of flavour. With that in mind, any potential change in our taste buds will naturally drive an altered perception even further, and so the future of our sense of taste deserves a segment of its own.

If you happen to be interested in history, you may at times have grimaced at the description of some of the foods our ancestors ate, from maggots to mice and whale gallstones.[212] It is quite tempting to suggest our sense of taste must have evolved in a rather short period of time, and therefore may be prone to future changes as well. Rather than having evolved, an alternative explanation for our altered food choices over time may be driven by the fact that our personal preferences adapt very quickly. Most adults likely eat some type of food that they as children swore they would never touch once they moved away from home. Alternatively you may be like me, and enjoyed liver pancakes throughout childhood, only to now turn green to its very scent a few decades later.

Again, in this example we are talking about flavour rather than true taste, though it is difficult to discuss one without the other. The hunt for flavour, or tastes, has after all been an incredibly important force for human development. Our desire to eat well has inspired a range of new inventions over the years, from farming techniques that have staved off hunger, to tools required for hunting and chemical techniques relating to preservation and safety. The development of umami has shown that the way we talk about taste is continuously changing, and plant-based meat is currently being developed in several parts of the world. It is consequently very clear that the keenness for exploring flavour inspires development, and development in turn changes our perspective on tastes.

Let us however look at the physiology, and consider the possible future evolution of our taste buds. Much like smell, there is some evidence for genetic predispositions to certain tastes which range between populations. A study looking into taste perceptions between Asian and Caucasian individuals in Britain found that the former were, on average, more sensitive to bitter and sour tastes.[213] On top of that, Asian participants were more sensitive to detecting the temperature of food and were found to enjoy sweet tastes less.

The earlier chapter on taste discussed a number of differences between cultures, and it is notable that the Asiatic region stood out as having a quite distinct relationship to taste and flavour. It may be unclear why our sense of taste has been subject to this very rapid evolution, and to what extent it reflects cultural or environmental factors. It does however show that taste, much like smell, can evolve in a rather short span of time, and we have clear examples of how different perceptions of taste have impacted the way we view and talk about it.

Without knowing the purpose for our deviating taste buds, it is difficult to predict what the future of taste may look like. It is quite clear that food is becoming increasingly global; as a child I had no idea what sushi was, yet today it is readily accessible in most western countries. Food fusion, the mixing of different cuisines, is also seeing a rise in popularity. In this way, it seems likely that the way forward will be one of gustatory unification as we are exposed to similar tastes, and talking about food in a more universal manner will likely also influence the way we perceive it.

For humanity, our sense of taste has in many ways already served its purpose in guiding us away from the potentially lethal and towards the edible; we rarely taste something to ensure its edibility, but instead trust our local food vendors and the best-before-dates. You may however suggest that being averse to sugary foods would be beneficial, as it may decrease mortality and increase fertility. Still, it is difficult to say if this would influence birth rates in any significant way, given that obesity is not a novel phenomenon and no such trend has been recorded on a global scale. Much like smell, it seems likely any inter-individual differences in perception of taste should diminish over time. In other words, we are likely growing closer with every newly opened restaurant.

The Future of Touch

It may seem unlikely that something as primordial as our sense of touch may go through any noticeable evolutionary changes in the near future. While our tools may have changed, almost all devices are ultimately designed to be compatible with a tactile user-interface. So what would be the driving force behind any further evolution in our sense of touch? As with most other senses, modern culture and technology have changed the way we view this essential form of interaction and communication.

Touch is critical for human well-being, both for mental and physical health. As our ancestors emerged from the jungles, social grooming was a centrepiece of life.[214] Even today, the absence of touch can contribute to depression and mental ill-health,[215] so what will happen if our society grows increasingly individualistic? Western cultures in particular are becoming increasingly deprived of touch; a study from the 1960's compared the interaction between two friends from different countries, and saw that British pairs would almost never touch each other during a conversation, while Puerto Ricans led the league with 180 physical interactions.[215] The benefits are well-documented and the science clear: touch releases oxytocin, a hormone contributing to the perception of happiness, and we even know that infants will exhibit healthier development if experiencing regular human touch.[215]

The phrase "use it or lose it" seems to hold true for many aspects of sensory neuroscience; we have discussed how several animals lost their taste for sweetness having had no use for it due to their feeding habits. Could our human sense of touch go the same way if not utilised? If we grow unhappy due to sharing fewer tactile interactions, surely it would be convenient if we stopped being so influenced by the experience. In the end, we must ask ourselves the same question as with all other senses: would an individual with no emotional response to touch be more likely to produce many offspring? Probably not; in fact, it feels somewhat counterproductive. Considering its evolutionary importance, our innate and physiological responses to touch are unlikely to become redundant.

With that being said, there is evidence that our sense of touch has changed over time. Archaeological findings have revealed that several aspects influencing our tactile perceptions have changed over the scope of a few 1,000 years. One factor contributing to this is our sheer size. Human were generally shorter hundreds of years ago; for instance, 600 years ago, they were roughly as tall as 10,000 years ago.[216] Go back 40,000 years however, and people were generally taller than we are today. Our size naturally affects our body perception, strength, and general constitution. Smaller body sizes generally having a greater hair follicle density[217] and therefore greater sensitivity to touch. How will this affect touch in the future? Many populations are increasing in size and height due to better nutrition, but there are few advantages to being large in modern society as labour has become less and less physically demanding. Our increasing sizes however, may end up influencing the way we perceive touch.

A more direct phenomenon related to touch is that of handedness. About 90% of humans are right handed, with the remaining 10% being left handed, as true ambidexterity is exceptionally rare.[218] It is still hotly debated why humans and other primates show such a clear preference for one hand, most notably the right. Studies have shown that right-handedness may prove more efficient in manual dexterity[219] and as most primates rely on highly dexterous tasks, this may be part of the explanation. By contrast, there is some evidence that left-handed individuals express greater creativity,[220] although studies on intelligence have revealed no clear correlation either way.[221] The preference for one hand over the other is due to the lateralisation of the human brain, which sees one cerebral hemisphere dominating in one field relative to the other side of the brain. For example, language and maths are generally processed in the left side of the brain while the right is more important for visuospatial functions.[222] Handedness may affect how influential these other neural processes are, as they all relate to this neural lateralisation.[222]

Handedness, a critical part of touch, may therefore have significant influences on our lives. So has this changed over time? While we seem to see more left-handed people in the world today, it is still too early to say whether this is due to an increase in the phenomenon, or whether this is simply due to children being allowed to use their preferred hand; throughout history, left-handedness has often been cruelly discouraged by the schooling system or other institutions. Should the larger number of south-paws be due to a real increase, it is tempting to suggest that we may look forward to a more creative world. One should however note that hand preference is curiously very weakly inherited[218] so the future of handedness is difficult to predict.

Ultimately, our sense of touch has gone through a surprising number of changes in adapting to our ever-changing world, and may continue to do so. Touch-deprivation seems a likely future for mankind, even in highly tactile communities it may be a struggle to mimic the tactile frequency enjoyed by our ancestors and genetic progenitors. Still, we know the importance of touch, and we can hope that the nearness of other people will become increasingly appreciated, and prioritised. As for its genetic evolution, there seems to be no clear indication that humanity is in for any concrete changes in our tactile prowess, except for perhaps a slight decrease in tactile acuity with increased body sizes. No matter the case, we would do well to remember the importance of human touch.

Robotics, AI, and the Unknown

It is impossible to envision the future of our senses without taking into account the development of bionic prosthetics. Far from the simple wooden contraptions of old, we now have artificial eyes, ears, and limbs that can detect, compute, and relay sensory information. Some of them are better than others, though few can be said to surpass our own biological hardware. At least, that seems to hold true for now, but what about the future? As humans and machines become increasingly intertwined, our perception of the world may very well be tied to our digital and mechanical constructs.

Time and time again we have seen how the animal world has preceded our artificial sensors by millennia. Today, humanity's drive for exploration has allowed us to construct devices that could outperform organic life in several fields. We can detect chemical compositions on other planets, map the oceans, and anticipate changes in the earth or atmosphere. For now, these devices represent expensive endeavours, but the costs for technologies are decreasing rapidly.

So what about harnessing the immense power of human creativity in enhancing our own relatively moderate sensory capacities? The idea itself is quite old, with bionic enhancements featured in popular culture since the turn of the 20th century.[223] But can we expect fiction to become reality, as has happened with so many other technological advances? Would this even be advisable, considering the innate relationship between our brains and sensory organs which millions of years of evolution have forged? We will most certainly try, and have been doing so with varying levels of success for some time. Let us now take a look at the ways we, as humans, may change the direction of vision, hearing, smell, taste, and touch.

Bionic Vision

Vision is likely the most difficult sense to replicate artificially or bionically, as the immense complexity of the system makes the introduction of artificial input challenging. Consider the optic nerve, which reliably transfers information from the retina in a reliable and organised way. Attaching a camera to this nerve may appear an ideal method for relaying visual data. An alternative may be starting from the other end, as visual images are construed by the brain itself, meaning that we may project the camera's contents directly onto the occipital cortex, making the eyes obsolete. Both of these approaches have yielded positive results in recent studies, though have not caught up with fiction quite yet.

Few sensory deficits have as profound effects as that of lost vision. A functional bionic ocular prosthetic, however rudimentary, will therefore be extremely useful. At their current state, both the retinal and cortical implants are capable of only allowing a wearer to discern simple shapes and not much more.[224,225] For someone with visual impairment or blindness, this small improvement could nevertheless signify a vast improvement in their quality of life.

In order to surpass human eyes, there are several steps which prosthetics need to overcome. The general principles for the two types of prosthetics rely on similar elements. A camera relays visual information to a number of electrodes, implanted at key regions of the neural system. The peripheral system, which aims to send information through the optic nerve, employs a small chip with microscopic spikes that can be inserted into the back of the retina. Generating electrical current based on the features of the video, the spikes then send their signals through the optic nerve via the person's own eye.

Such an approach does of course only work if the person has an intact, functioning eye to begin with. The other option, which involves a surgeon place stimulation electrodes in the visual cortex, foregoes any such requirements. While the procedure is more invasive, it bypasses any potential issues arising from ocular disorders. Much like the retinal electrodes however, the wearer is only able to view very rudimentary figures. The limited number of electrodes that one may place in the eye or brain is a natural obstacle and we may struggle to ever be able to stimulate the brain with the spatial precision needed to mimic organic vision, as we cannot place an electrode for every individual neuron. The stimulation method instead relies on creating phosphines, so a person sees light without any having entered the eye, much like you may have experienced after looking into the sun for too long. The image viewed through these prosthetics is at present little more than patterns formed by these phosphines.

Bionic eyes therefore seem to have a long way to go before they can compete with biological vision. Not only does the spatial resolution need improvement, but video tracking of objects still requires quite a bit of work, as we have seen in the field of self-driving cars. Beyond using prosthetics to help those with visual impairment regain some visual function, it is also worth considering whether prosthetics could ever be developed for visual enhancement in those with normal vision. For instance, could we use a similar setup as the ones above to relay information about infrared or ultraviolet radiation directly into the mind of a wearer in the form of phosphines? It seems unlike that this would be any real benefit, as eyewear and handheld tools can provide much clearer information without the need for invasive surgery. Instead, the future of visual aids may lie more in the realm of augmented reality, simply providing data through wearable high-tech spectacles.

Bionic Hearing

Hearing has been subject to technological augmentation for over a 100 years. Alexander Graham Bell's invention of the telephone in 1876 meant that sound could be modified through mechanical means, and it was in the 1920s that hearing aids featuring vacuum tubes capable of transforming speech into electrical signals first appeared in a wearable device. Today, hearing aids are far more advanced, and may well hold an interesting role for the future of our hearing.

The principle behind the hearing aid, the most common form of bionic hearing, is much the same as it was a 100 years ago: a microphone picks up the sound, it is transformed into digital signals, and finally the signal is once again transmitted as sound through a microphone.[226] Systems like these can be found in almost any modern headset, and play an important role in several industries where communication is needed. Still, individuals who suffer from conductive hearing loss cannot benefit from such systems, as the sound cannot reach the inner ear because of damages to the conductive bones. Bone conduction, like that implemented by Beethoven and where a sound is relayed to the cochlea through the skull rather than the ear, may serve as an alternative.

This technology is becoming increasingly viable for other uses as well. Bone-conductive earphones are available for commercial use and can allow you to listen to sound while under water, since the device does not require any medium other than bone to convey its signal. Companies are also working on wearable devices that may allow you to communicate through bone conduction, such as rings that send a signal through a wearer's index finger, so that when it is placed next to the ear the signal may be heard as if through a headset. We may therefore expect bone-conductivity to play a bigger role in the future of hearing, both clinically and recreationally.

A more interesting aspect of hearing aids may be the potential to augment the neural signal itself. Cochlear implants involve five distinct steps: first, a microphone detects a sound, a processor then translates the signal into electrical signals, a transmitter send this information to a receiver in the ear, which finally relays the impulse through a number of electrodes to the auditory nerves.[226] Such systems have shown promise in allowing patients with severe sensory hearing loss to regain a functioning level of speech perception. Again, the number of electrodes or channels available influences the quality of the signal, as does the algorithm that translates the sound. In terms of efficacy, cochlear implants may allow 60–70% of a sentence to be viably understood, though results vary greatly between people and the greatest benefit for the user is generally an improved sound localisation. A final type of hearing prosthetic involves stimulating the brainstem directly, bypassing the auditory nerves. While some people have regained a certain level of hearing perception following brainstem stimulation, outcomes are primarily limited to enhancing the perception of loud sounds, like car horns or alarms, which may of course be life-saving in many situations.[227]

The prospect of bionic hearing becoming more sophisticated appears to be quite promising. We use the technology regularly in our everyday lives, and merging brain and machine offers benefit for those with hearing impairment. Even those with normal hearing function may be able to benefit from auditory technology, with the absence of soundwaves in bone conductivity meaning that a message would be very difficult to eavesdrop on without analysing the vibrations of the listener's skull. It also ignores whatever medium the ear is exposed to, and could prove a valuable asset for divers. So while bionic hearing is and old phenomenon, it will likely continue to develop as our need to communicate across distances remains a key feature in our ever-growing societies.

Bionic Smell

Relatively little research has been carried out on functional smelling prosthetics, at least compared to those for vision and hearing.[228] That is not to say that the field is not important, as we have seen in the previous chapter on smell. Perhaps part of the reason for the relative lack of focus on developing technology for smell is that the sense does not play as important a role in communication as either

Nevertheless, artificial noses are a reality, and these have to some extent already begun to be merged with a human host.

Operating according to the same general principles as a cochlear implant, olfactory prosthetics may rely on external sensors which would be able to detect odour molecules, process and then transmit this information as digital data to a set of stimulation electrodes in the olfactory epithelium.[228] While the sensors exist, the problem of translating them to the brain in a meaningful way remains. Much like cochlear implants need to be programmed for the individual, with the signal being fine-tuned and adjusted until it becomes perceptible, a smelling device would also have to be set up for each individual. It appears likely that this experience would not be altogether pleasant, considering how the brain tends to interpret foreign or disparate smells as abrasive. Despite this risk, considering the evidence for similar setups working for hearing, it appears quite likely that unique patterns can be constructed by processing algorithms allowing certain scents to be detected like olfactory fingerprints.

With quite little resources being devoted to this line of research, it may be several years before we see viable olfactory prosthetics develop. While the prototypes appear highly likely to take shape in the future, it does seem unlikely that we shall see them anytime soon. On the other hand, the loss of smell caused by the Covid pandemic has renewed interest in this field, and so only time will tell what the next generation of engineers and entrepreneurs may choose to prioritise in our changing medical landscape.

Bionic Taste

Taste has experienced the least amount of technological progress of all the senses, at least when it comes to machine-brain-interfacing. Artificial tongues are already capable of detecting and processing complex tastes,[229] so it may appear that the framework for prosthetic use would exist. However, tongues are quite complex, relying on several cranial nerves to produce a range of movements involved in eating or speaking.[19] While taste is certainly an important sense, focus has been on developing prosthetics capable of allowing these important motor functions over any gustatory capacity. As we saw in the previous chapter on taste, taste buds are not limited to the tongue, even though this is where most of them are concentrated. We are therefore rarely completely left without taste, unless affected by a disorder influencing the central relay or processing of the sensory signal.

With no current bionic taste examples, it is difficult to anticipate how bionics may play a part of its future. However, much as is the case with smell, the Covid-19 virus has left many people with an altered sense of taste, often long after the acute infection has resolved and we may see an increased demand for gustatory aid in the years to come. It is tempting to imagine how electrodes implanted into the brain could provide us with programmable taste experiences without the need to consume any sugary or fatty foods. While such solutions may be beneficial to our general health, it does appear a great extravagance, and there is currently no indication that such demands for neurosurgical procedures are gaining a foothold. Taste will probably not receive its bionic makeover until substantial progress has been made in the field, likely after similar protocols for our other senses have been clinically established. Until this turning point, the future of taste appears to rest with Mother Nature.

Bionic Touch

Artificial limbs may likely be what first cross one's mind when talking about bionic prosthetics. Much like hearing aids, mechanical limbs with sensory feedback have been around for over a 100 years; the movement of an artificial limb was relayed to healthy structures through pneumatic pressure.[230] For a long time, these haptic sensory feedback systems relied on signalling to nearby limbs. Using visual cues, the wearer would gradually come to associate this tactile feedback with the action of the prosthetic and, thanks to the adaptive brain, manage to gain some sense of agency from the artificial structure. Supplementary feedback of this kind continues to be an important aspect of bionic touch, but much like for our other senses, the future of touch seems to lie with the merging of brain and machine.

Touch is old, at least biologically speaking. Considering its exceptional importance, it may therefore come as a surprise that we still have much to learn about this most primordial sense. It was only in 2021 that the Nobel Prize was awarded to David Julius and Ardem Pataputian for their discoveries on how the brain can detect touch through peripheral receptors.[231] These findings are bound to inspire and guide a whole new line of research into artificial touch.

Artificially, we still struggle to mimic the precision of human movements or touch. Nerves and muscles operate according to highly mechanical principles, which we know quite a bit about. We have mapped which nerves innervate each part of the hand and where in the somatosensory cortex signals related to touch are perceived. Despite this insight, providing bionic limbs with sensory input as well as synchronised motor control has proven challenging; focus has historically been on improving on the latter aspect, so that a person may use a bionic limb for grasping or other important tasks. It is however clear that in order to achieve greater agency of a new limb, the sense of touch plays a crucial role. Without it, the brain will lack any proprioceptive input to coordinate the movement, and will have to rely on secondary visual information to adjust its commands.

An integrated sensorimotor prosthetic is indeed already a reality. Three Swedish patients have been living with sensing artificial arms for several years.[232] These self-contained devices do not rely on any external power sources or medical supervision and they allow its wearers a ground-breaking level of autonomy. What makes these arms so special is that all necessary computations are carried out in the periphery, in the arm itself. The prosthetic is attached to the person's bone, acting as a natural elongation of the limb, and is surgically grafted with nerves and muscles. Motor nerve activity, as produced by the owner's brain, is translated into the arm's own processor, which leads to an appropriate movement. Pressure sensors on the thumb collect touch information which is carried back through the limb, before being processed and relayed through the person's own sensory nervous system. In this way, a person can decide to grasp an object, receive feedback from their action, and adjust accordingly. With data from this study being published in 2020, the method is very new. The intervention may well prove to be a crucial stepping stone for the future of bionic limbs.

From the perspective of artificial touch, the greatest obstacle is likely acquiring the same level of sensitivity as a biological structure. In the artificial arms described above, the pressure sensor on the thumb can provide very rudimentary, albeit highly significant, sensory input. In order to acquire greater sensitivity, more invasive procedures may be needed. The somatosensory cortex is quite substantial, and several regions are available from the surface of the brain. Studies have been able to produce sense perceptions of the hand by stimulating these accessible areas, but more delicate sensory inputs have been difficult to achieve due to the structures that normally process them being hidden away in the deeper, sulcal, areas.[233] Using a new type of electrode, neurosurgeons have very recently been able to reach these areas in two patients. This could very well be a turning point for bionic touch, as it may finally allow that highly localised sensory input that our fingers provide, crucial for tasks requiring fine motor skills. Still, with invasive brain surgery being less accessible than peripheral solutions, and these technologies being quite new, our organic limbs still hold the figurative upper hand.

The combination of motor and sensory systems in a self-controlled prosthetic does invite speculation for what the future may hold in terms of touch. Brain implants in the motor cortex of monkeys have allowed them to move objects on a computer screen.[234] While this technology is a few decades old, it has recently been increasingly popularised by commercial companies, such as Elon Musk's Neuralink, which have been able to repeat these trials with a comparatively high electrode count.[235] Considering that a bionic limb is just the physical representation of digital processes, there is little to stop us from issuing motor commands, and receiving sensory feedback, in a completely digital medium. A central computer program could then repeat the activity for any number of real-life scenarios. For example, a highly skilled surgeon may in the future not only control surgical equipment, they may be able to feel them as if they were actively touching the scalpels. Robots may be programmed to move in tandem with a human controller, obeying its commands and relaying its sensory input from possibly dangerous locations. Whole-body prosthetics of this kind would also invite the question of agency and consciousness being transferred from man to machine. As we previously discussed, human nerves are relatively slow compared to electrical wiring, and artificial nerves are already being produced.[236] It may therefore not be unthinkable that those who are unable to move or touch due to paralysis may in the future not only benefit from artificial limbs, but even inhabit prosthetic bodies.

Touch does indeed offer some exciting speculation on what is yet to come for sensory neuroscience, with both technology and popular interest pushing the advancement. Out of all artificial systems, I imagine touch may be the first to surpass the biological form. While the current technology is a far cry from the cybernetic future we see in popular media, the future of touch may well be one of electrical wiring and artificial enhancements.

Artificial Intelligence and the Unknown

The goal of this book is to relay facts rather than to speculate. As this chapter has shown, such an ambition was difficult to keep. There is simply too much research being carried out in the field to completely forego the potential future state of our senses. It may be disappointing that we are still far from achieving artificial constructs that may compete with our biological sensors, but progress in the field of bionics may nevertheless give us greater insight into one of the great unknowns of life, consciousness.

We previously dealt with consciousness as a potential internal sense, closely related to agency and free will. Scientists have made significant headway into exploring how our sensory inputs guide the perception of our own bodies, and how agency can be transferred to artificial limbs through a combination of tactile and visual cues.[237] Considering the close relationship between sensory input and our idea of who we are, sensing machines become increasingly interesting for answering some of life's great questions.

It has been argued that multisensory integration, linking all our perceptions into a unified experience, represents a "unity of consciousness".[238] With exponential improvement in artificial intelligence and machine learning, it is tempting to suggest that truly intelligent machines, capable of independent thought, may also rely on a conglomeration of complex sensory inputs. *Neural correlates of consciousness* make up a scientific theory of how consciousness is achieved through a minimal level of neural mechanisms and events.[239] If we are able to construct artificial limbs capable of detecting and integrating sensory information, where do we stand on building a prosthetic consciousness?

Artificial intelligence aims to mirror or even surpass humans within a predetermined field. Definitions of the term differ between scholars, but a broad understanding is that AI should be capable of allowing a machine to interpret its surroundings so that it may take appropriate actions in realising a predetermined goal.[240] It is therefore clear that sensory inputs play a crucial role in this field of research. If consciousness emerges from multisensory integration, does that mean that a complex enough computer, fed with enough sensory data, may one day become sentient?

This aspect of AI research deals with what is called *general* artificial intelligence. Specific purpose algorithms have seen a tremendous increase in complexity over the last few years.[241] While these machines are capable of incredibly complex work, they cannot think for themselves, and may perhaps learn in the process of carrying out their tasks. They are nevertheless limited to their own predetermined coding. Sentient machines would rely on general AI, which could mimic the human brain's level of complexity for problem solving and planning. Some experts claim that we have a 50% chance of achieving this milestone before 2060,[242] while others say the task is impossible.[241] A major obstacle in this pursuit is that we are still unclear on how the neural correlates of consciousness are achieved, but it has been suggested that quantum computing, carried out by the organic computer that is our brain, plays an important role.[243] With progress being made in quantum computing, taking advantage of complex physical properties of subatomic particles, we may soon be in a better position to see where this trajectory will lead us.

The importance of sensory neuroscience on these systems should not be underestimated. Imagine a thinking robot, capable of guiding you through life, but incapable of perceiving the world in the same way you do. Things we take for granted may fall within a small margin for error within the sentient machine, and complications may be devastating. On the other hand, fitting these systems with superhuman sensors, as I believe we are bound to do, may make them appear near supernatural to us and create an inevitable synthetic-organic divide. Some may hope that we never achieve general AI for these very reasons, but our attempts to do so will almost certainly aid us in the pursuit of understanding our own consciousness. With it, our perceptions of the world will undoubtedly continue to adapt. The future will be an exciting time for understanding, improving, and implementing our senses, though we can only speculate on what shape or form they may take.

Chapter 11

A Final Remark

Throughout these chapters, we have explored the non-human senses of detecting magnetic and electric fields, including how arachnids use the latter for aerial assaults, the sensory affinities for fine changes in atmospheric pressure, radioactive radiation, photon spin in polarised light, moisture, fires, balance through interpreting light signals, and the tardigrade's disregard for sensory systems other than touch. Humans might not be able to sniff out radiation and the only way we may perceive Earth's electromagnetic field is through the astonishing sights of the Northern Lights. Our senses are much more universal than the highly specialised physiologies of the bat, dog or dolphin, and this is a reflection of our ability to exist in a wide variety of landscapes. On top of this generalist platform on which we humans stand, the steep upward curve of technological advancements will undoubtedly open up new and exciting realms for our senses to explore. So do not let this sensory variety seen in the animal kingdom bring you down. We are exactly where we need to be, so let us instead find solace in our mediocracy, and enjoy the fantastic story of life on Earth as told by the species that inhabit it.

Vision, hearing, smell, taste and touch make up the basis for the way we perceive the world. They also reflect the way we think about it, having grown from a wide variety of inward and outward wits into a distinct Five Senses as human culture evolved. There certainly are other senses, both external and internal, and recognising them for what they are is hugely important as all aspects of life are experienced in reference to these sensory appendages of the brain. As technology continues to evolve we may end up being able to detect aspects of our universe we cannot even conceptualise today, like radioactivity would have appeared to the Vikings, but those findings will always have to be put into a context which we can understand.

I am hoping that this book shed a little light into the hidden mechanisms that allow us to experience the world and each other. One could imagine that the brain works in darkness, only lit up by the wandering fairy lights of our neural impulses as they enter beneath the folds. Perhaps reading some of these words has challenged that notion, because we do not hear with our ears and nor do we smell with our nose. We hear, smell, see, taste and touch with the brain, and what we call sensory organs are just the means through which the information gets there.

Human beings are complex. I would never ask that someone remembers the details of the visual system after having read such a brief description as the one presented here. What I do hope however is that it provided a little bit of insight into the physiological realm of the senses. Considering the improbability of life, it is astonishing how evolution has found a way of making every aspect of this planet work with its inhabitants. The more one studies the biological mechanisms that allow this symbiosis to thrive, the more one realises that we really know very little. Today we might think of the ancient description of the senses as something funny, making references to wind, earth, water and fire, not to mention Descartes who made our neural network into a pneumatic tubing system. We should be prepared that future generations may say the same things of us, wondering how we got on knowing so little about the very systems that make us who we are.

I have heard it being said in crestfallen tones that we are born into a generation too late to explore the Earth, and too early to cruise the stars. This is an unnecessarily gloomy perspective of life. At no time throughout history has there ever been a better time for discovering the very essence of our own existence; we are born into a generation beautifully aligned for exploring the human mind. Never before have such vast amounts of knowledge been available to so many, and it will not be technology itself that solves the mysteries of existence but ordinary human beings like you and me. I might at times have alluded to comparing our neural system to that of machinery, with electrical impulses and on-off switches, but this is because art mimics reality. There is so much more out there to discover, and I am confident that together we can find a way to make some sense out of it all.

References

1. Kandel, E. R., Schwartz, J. H., & Jessell, T. M. *Principles of Neural Science* (4th edn.) (New York: McGraw-Hill, 2000).
2. Anderson, E. R. *Folk-taxonomies in Early English.* (Fairleigh Dickinson Univ Press, 2003).
3. White, K. & Macierowski, E. M. *Commentaries on Aristotle's" On Sense and What Is Sensed" and" On Memory and Recollection"(Thomas Aquinas in Translation).* (Catholic University of America Press, 2005).
4. Bakalis, N. *Handbook of Greek Philosophy: From Thales to the Stoics Analysis and Fragments* (Trafford on Demand Pub, 2005).
5. Hume, R. E. *The Thirteen Principal Upanishads: Translated from the Sanskrit with an Outline of the Philosophy of the Upanishads and an Annotated Bibliography* (H. Milford: Oxford University Press, 1921).
6. Milner, M. *The Senses and the English Reformation* (Routledge, 2016).
7. Frantzen, A. J. *Bloody Good: Chivalry, Sacrifice, and the Great War* (University of Chicago Press, 2004).
8. Postgate, J. R. *The Outer Reaches of Life* (Cambridge University Press, 1994).
9. Hicks, R. D. *Aristotle De Anima* (Cambridge University Press, 2015).
10. Kolve, V. A. *Chaucer and the Imagery of Narrative: The First Five Canterbury Tales* (Stanford University Press, 1984).
11. Barber, C. L. *Early Modern English* (Edinburgh University Press, 1997).
12. Nordenfalk, C. The five senses in late medieval and Renaissance art. *Journal of the Warburg and Courtauld Institutes* **48**, 1–22 (1985).
13. Land, M. F. & Fernald, R. D. The evolution of eyes. *Annual Review of Neuroscience* **15**, 1–29 (1992).

14. Schoenemann, B., Pärnaste, H., & Clarkson, E. N. K. Structure and function of a compound eye, more than half a billion years old. *Proceedings of the National Academy of Sciences* **114**, 13489, doi:10.1073/pnas.1716824114 (2017).

15. Feuda, R., Hamilton, S. C., McInerney, J. O., & Pisani, D. Metazoan opsin evolution reveals a simple route to animal vision. *Proceedings of the National Academy of Sciences* **109**, 18868–18872 (2012).

16. Rothman, L. S. *et al.* The HITRAN 2004 molecular spectroscopic data-base. *Journal of Quantitative Spectroscopy and Radiative Transfer* **96**, 139–204 (2005).

17. Snell, R. S. & Lemp, M. A. *Clinical Anatomy of the Eye* (John Wiley & Sons, 2013).

18. Shaw, G. M. *et al.* Epidemiologic characteristics of anophthalmia and bilateral microphthalmia among 2.5 million births in California, 1989–1997. *American Journal of Medical Genetics Part A* **137**, 36–40 (2005).

19. Purves, D. E. *et al.* Neuroscience. (2008).

20. Horton, J. C. & Adams, D. L. The cortical column: A structure without a function. *Philosophical Transactions of the Royal Society B: Biological Sciences* **360**, 837–862 (2005).

21. Markowsky, G. *Encyclopædia Britannica*, https://www.britannica.com/science/information-theory (2017).

22. Vatansever, D., Menon, D. K., & Stamatakis, E. A. Default mode con-tributions to automated information processing. *Proceedings of the National Academy of Sciences* **114**, 12821–12826 (2017).

23. Servos, P. & Goodale, M. A. Preserved visual imagery in visual form agnosia. *Neuropsychologia* **33**, 1383–1394 (1995).

24. Turnbull, C. *The Forest People* (Random House, 2015).

25. Ferreira, C. T., Ceccaldi, M., Giusiano, B., & Poncet, M. Separate vis-ual pathways for perception of actions and objects: Evidence from a case of apperceptive agnosia. *Journal of Neurology, Neurosurgery & Psychiatry* **65**, 382–385 (1998).

26. Rubens, A. B. & Benson, D. F. Associative visual agnosia. *Archives of Neurology* **24**, 305–316 (1971).

27. Damasio, A. R. Prosopagnosia. *Trends in Neurosciences* **8**, 132–135 (1985).

28. Rossing, T. D. & Stumpf, F. B. The science of sound. *American Journal of Physics* **50**, 955–955 (1982).

29. D'Ambrose, C. Frequency range of human hearing. *The Physics Factbook* (2003).

30. Clack, J. A. Evolutionary biology: The origin of terrestrial hearing. *Nature* **519**, 168 (2015).
31. Quam, R. *et al.* Early hominin auditory capacities. *Science Advances* **1**, e1500355 (2015).
32. Bolhuis, J. J., Tattersall, I., Chomsky, N., & Berwick, R. C. How could language have evolved? *PLoS Biology* **12**, e1001934 (2014).
33. Wade, N. *In Click Languages, an Echo of the Tongues of the Ancients*, https://www.nytimes.com/2003/03/18/science/in-click-languages-an-echo-of-the-tongues-of-the-ancients.html (2003).
34. Netter, F. H. & Colacino, S. *Atlas of Human Anatomy*. (Ciba-Geigy Corporation, 1989).
35. Akeroyd, M. A. The psychoacoustics of binaural hearing: La psicoacústica de la audición binaural. *International Journal of Audiology* **45**, 25–33 (2006).
36. Crich, T. *Recording Tips for Engineers: For Cleaner, Brighter Tracks* (Routledge, 2017).
37. Bansal, M. *Essentials of Ear, Nose & Throat* (JP Medical Ltd, 2016).
38. Ramirez-Moreno, D. F. & Sejnowski, T. J. A computational model for the modulation of the prepulse inhibition of the acoustic startle reflex. *Biological Cybernetics* **106**, 169–176 (2012).
39. Stevens, K. M. & Hemenway, W. G. Beethoven's deafness. *Jama* **213**, 434–437 (1970).
40. Yongbing, S. & Martin, W. Spontaneous otoacoustic emissions in tinnitus patients. *Journal of Otology* **1**, 35–39 (2006).
41. Cazals, Y. Auditory sensori-neural alterations induced by salicylate. *Progress in Neurobiology* **62**, 583–631 (2000).
42. Hobson, J., Chisholm, E., & El Refaie, A. Sound therapy (masking) in the management of tinnitus in adults. *Cochrane Database of Systematic Reviews* (2012).
43. Kish, D. Human echolocation: How to "see" like a bat. *New Scientist* **202**, 31–33 (2009).
44. Downey, G. Sensory enculturation and neuroanthropology: The case of human echolocation. In *The Oxford Handbook of Cultural Neuroscience* (pp. 41–55) (Oxford University Press, 2016).
45. Kremer, W. Human echolocation: Using tongue-clicks to navigate the world. *BBC World Service*, http://www.bbc.com/news/magazine-19524962 (参照 2015-01-15) (2012).
46. Currier, K. E. *Auditory Representation in Spatial Applications*, University of California, Santa Barbara, (2011).

47. Zimmer, C. *The Smell of Evolution*, https://www.nationalgeographic.com/science/phenomena/2013/12/11/the-smell-of-evolution/ (2013).

48. Niimura, Y. Olfactory receptor multigene family in vertebrates: From the viewpoint of evolutionary genomics. *Current Genomics* **13**, 103–114 (2012).

49. Malnic, B., Godfrey, P. A., & Buck, L. B. The human olfactory receptor gene family. *Proceedings of the National Academy of Sciences* **101**, 2584–2589 (2004).

50. Nieuwenhuys, R., Voogd, J., & Van Huijzen, C. *The Human Central Nervous System: A Synopsis and Atlas* (Springer Science & Business Media, 2007).

51. Carskadon, M. A. & Herz, R. S. Minimal olfactory perception during sleep: Why odor alarms will not work for humans. *Sleep* **27**, 402–405 (2004).

52. Keller, A., Zhuang, H., Chi, Q., Vosshall, L. B., & Matsunami, H. Genetic variation in a human odorant receptor alters odour perception. *Nature* **449**, 468 (2007).

53. Doty, R. L. & Cameron, E. L. Sex differences and reproductive hormone influences on human odor perception. *Physiology & Behavior* **97**, 213–228 (2009).

54. Goetz, C. G. *Textbook of Clinical Neurology* (Vol. 355) (Elsevier Health Sciences, 2007).

55. Leopold, D. & Meyerrose, G. In *Olfaction and Taste XI* 618–622 (Springer, 1994).

56. Leopold, D. Distortion of olfactory perception: Diagnosis and treatment. *Chemical Senses* **27**, 611–615 (2002).

57. Hawkes, C. H. *Smell and Taste Complaints* (Butterworth-Heinemann, 2002).

58. Frasnelli, J. *et al.* Clinical presentation of qualitative olfactory dysfunction. *European Archives of Oto-Rhino-Laryngology and Head & Neck* **261**, 411–415 (2004).

59. Landis, B., Frasnelli, J. & Hummel, T. Euosmia: A rare form of parosmia. *Acta Oto-Laryngologica* **126**, 101–103 (2006).

60. Röck, F., Barsan, N., & Weimar, U. Electronic nose: Current status and future trends. *Chemical Reviews* **108**, 705–725 (2008).

61. Mischley, L. & Rountree, R. Preventing and slowing the progression of parkinson's: A clinical conversation with laurie mischley, ND, MPH, PhD, and Robert Rountree, MD. *Alternative and Complementary Therapies* **25**, 59–67 (2019).

62. George, A. The woman who can smell Parkinson's. *New Scientist* **241**, 40–41 (2019).
63. Kielan-Jaworowska, Z., Cifelli, R. L., & Luo, Z.-X. *Mammals from the Age of Dinosaurs: Origins, Evolution, and Structure* (Columbia University Press, 2004).
64. Yarmolinsky, D. A., Zuker, C. S., & Ryba, N. J. Common sense about taste: From mammals to insects. *Cell* **139**, 234–244, doi:10.1016/j.cell.2009.10.001 (2009).
65. Breslin, P. A. An evolutionary perspective on food and human taste. *Current Biology* **23**, R409–R418 (2013).
66. Lindemann, B., Ogiwara, Y., & Ninomiya, Y. The discovery of umami. *Chemical Senses* **27**, 843–844 (2002).
67. Pelletier, C. Beyond the tongue map: Evaluating taste and smell perception. *The ASHA Leader* **7**, 6–20 (2002).
68. Spence, C. Why is piquant/spicy food so popular? *International Journal of Gastronomy and Food Science* **12**, 16–21 (2018).
69. Hedayat, K. M. & Lapraz, J.-C. In *The Theory of Endobiogeny*, Kamyar M. Hedayat & Jean-Claude Lapraz (eds.) (pp. 185–213) (Academic Press, 2019).
70. Hadhazy, A. *Tip of the Tongue: Humans May Taste at Least 6 Flavors*, https://www.livescience.com/17684-sixth-basic-taste.html (2011).
71. Bartoshuk, L. M., Duffy, V. B., Hayes, J. E., Moskowitz, H. R., & Snyder, D. J. Psychophysics of sweet and fat perception in obesity: Problems, solutions and new perspectives. *Philosophical Transactions of the Royal Society B: Biological Sciences* **361**, 1137–1148 (2006).
72. Wang, Y. Y., Chang, R. B., & Liman, E. R. TRPA1 is a component of the nociceptive response to CO2. *Journal of Neuroscience* **30**, 12958–12963 (2010).
73. Graber, M. & Kelleher, S. Side effects of acetazolamide: The champagne blues. *The American Journal of Medicine* **84**, 979–980 (1988).
74. Bremer, J. In *Forum Philosophicum* (pp. 73–87).
75. Comer, C. & Baba, Y. Active touch in orthopteroid insects: Behaviours, multisensory substrates and evolution. *Philosophical Transactions of the Royal Society B: Biological Sciences* **366**, 3006–3015 (2011).
76. Schütz, C. & Dürr, V. Active tactile exploration for adaptive locomotion in the stick insect. *Philosophical Transactions of the Royal Society B: Biological Sciences* **366**, 2996–3005 (2011).
77. Dürr, T. P. V. The World of Touch. *Scholarpedia* **10**, 32688, doi:10.4249/scholarpedia.32688 (2015).

78. O'Madagain, C., Kachel, G., & Strickland, B. The origin of pointing: Evidence for the touch hypothesis. *Science Advances* **5**, eaav2558, doi:10.1126/sciadv.aav2558 (2019).

79. Kaas, J. H. The evolution of the complex sensory and motor systems of the human brain. *Brain Research Bulletin* **75**, 384–390 (2008).

80. Franzén, O., Johansson, R., & Terenius, L. *Somesthesis and the Neurobiology of the Somatosensory Cortex* (Springer Science & Business Media, 1996).

81. McGowan, C., Goff, L., & Stubbs, N. Animal physiotherapy. *Assesment, Treatment and Rehabilitation of Animals JAV* (Blackwell Publishing, 2007).

82. Furman, J. M. & Cass, S. P. Benign paroxysmal positional vertigo. *New England Journal of Medicine* **341**, 1590–1596 (1999).

83. Leigh, R. J. & Zee, D. S. *The Neurology of Eye Movements* (Oxford University Press, USA, 2015).

84. Sherrington, C. (Oxford University Press, 1906).

85. Truini, A., Barbanti, P., Pozzilli, C., & Cruccu, G. A mechanism-based classification of pain in multiple sclerosis. *Journal of Neurology* **260**, 351–367, doi:10.1007/s00415-012-6579-2 (2013).

86. Bautista, D. M. *et al.* TRPA1 mediates the inflammatory actions of environmental irritants and proalgesic agents. *Cell* **124**, 1269–1282 (2006).

87. Indo, Y. *et al.* Mutations in the TRKA/NGF receptor gene in patients with congenital insensitivity to pain with anhidrosis. *Nature Genetics* **13**, 485 (1996).

88. Castro, J. *11 Surprising Facts about the Circulatory System*, https://www.livescience.com/39925-circulatory-system-facts-surprising.html (2013).

89. Hammel, H. T. & Pierce, J. Regulation of internal body temperature. *Annual Review of Physiology* **30**, 641–710 (1968).

90. Kraft, S. *Everything You Need to Know About Hypothermia*, https://www.medicalnewstoday.com/articles/182197.php (2018).

91. Turk, E. E. Hypothermia. *Forensic Science, Medicine, and Pathology* **6**, 106–115 (2010).

92. Barnes, J. Posterior analytics (1994).

93. Gregoric, P. & Pavel, G. *Aristotle on the Common Sense* (Oxford University Press, 2007).

94. Lokhorst, G.-J. In *The Stanford Encyclopedia of Philosophy*, Edward N. Zalta (ed.) (2018).

95. Libet, B., Gleason, C. A., Wright, E. W., & Pearl, D. K. *Neurophysiology of Consciousness* (pp. 249–268 (Springer, 1993).

96. Schurger, A., Sitt, J. D., & Dehaene, S. An accumulator model for spontaneous neural activity prior to self-initiated movement. *Proceedings of the National Academy of Sciences* **109**, E2904–E2913 (2012).

97. Zinner, S. H. Tourette disorder. *Pediatrics in Review* **21**, 372–383 (2000).

98. Kayser, A. S., Sun, F. T., & D'Esposito, M. A comparison of Granger causality and coherency in fMRI-based analysis of the motor system. *Human Brain Mapping* **30**, 3475–3494 (2009).

99. Arain, M. *et al.* Maturation of the adolescent brain. *Neuropsychiatric Disease and Treatment* **9**, 449 (2013).

100. Parker, K. L., Lamichhane, D., Caetano, M. S., & Narayanan, N. S. Executive dysfunction in Parkinson's disease and timing deficits. *Frontiers in Integrative Neuroscience* **7**, 75 (2013).

101. Rao, S. M., Mayer, A. R., & Harrington, D. L. The evolution of brain activation during temporal processing. *Nature Neuroscience* **4**, 317 (2001).

102. Kerns, K. A., McInerney, R. J., & Wilde, N. J. Time reproduction, working memory, and behavioral inhibition in children with ADHD. *Child Neuropsychology* **7**, 21–31 (2001).

103. Mandler, G. Recognizing: The judgment of previous occurrence. *Psychological Review* **87**, 252 (1980).

104. Kinnavane, L., Amin, E., Olarte-Sánchez, C. M., & Aggleton, J. P. Detecting and discriminating novel objects: The impact of perirhinal cortex disconnection on hippocampal activity patterns. *Hippocampus* **26**, 1393–1413 (2016).

105. Jeannerod, M. The sense of agency and its disturbances in schizophrenia: A reappraisal. *Experimental Brain Research* **192**, 527 (2009).

106. Wegner, D. M. & Wheatley, T. Apparent mental causation. Sources of the experience of will. *The American Psychologist* **54**, 480–492, doi:10.1037//0003-066x.54.7.480 (1999).

107. Buchholz, V. N., David, N., Sengelmann, M., & Engel, A. K. Belief of agency changes dynamics in sensorimotor networks. *Scientific Reports* **9**, 1995, doi:10.1038/s41598-018-37912-w (2019).

108. Feinberg, T. E., Venneri, A., Simone, A. M., Fan, Y., & Northoff, G. The neuroanatomy of asomatognosia and somatoparaphrenia. *Journal of Neurology, Neurosurgery & Psychiatry* **81**, 276–281 (2010).

109. Moro, V., Pernigo, S., Zapparoli, P., Cordioli, Z., & Aglioti, S. M. Phenomenology and neural correlates of implicit and emergent motor awareness in patients with anosognosia for hemiplegia. *Behavioural Brain Research* **225**, 259–269 (2011).

110. Fox, S. I. *Human Physiology* (9th edn.) (New York, USA: McGraw-Hill press, 2006).

111. Hanke, F. D. & Osorio, D. C. Vision in cephalopods. *Frontiers in Physiology* **9**, 18 (2018).

112. Browers, L. Animal Vision Evolved 700 Million Years Ago. *Scientific American Blogs* (2012).

113. Ahnelt, P., Schubert, C., Kübber Heiss, A., & Anger, E. Adaptive design in retinal cone topographies of the Cheetah and other Felids. *Investigative Ophthalmology and Visual Science*, 195 (2005).

114. Thoen, H. H., How, M. J., Chiou, T.-H., & Marshall, J. A Different form of color vision in mantis shrimp. *Science* **343**, 411–413, doi:10.1126/science.1245824 (2014).

115. Land, M. *Encyclopædia Britannica*, https://www.britannica.com/science/photoreception (2019).

116. Calderone, J. *Here's Why Cats Have Such Strange, Haunting Eyes, Explained by Science*, https://www.sciencealert.com/here-s-why-cats-have-such-weird-eyes (2018).

117. Agi, E. *et al.* The evolution and development of neural superposition. *Journal of Neurogenetics* **28**, 216–232, doi:10.3109/01677063.2014.922557 (2014).

118. Miorelli, N. *Through the Compound Eye*, <https://askentomologists.com/2015/02/25/through-the-compound-eye/> (2015).

119. Borst, A., Haag, J., & Reiff, D. F. Fly motion vision. *Annual Review of Neuroscience* **33**, 49–70 (2010).

120. Barth, F. G. *A Spider's World: Senses and Behavior* (Springer Science & Business Media, 2013).

121. Forster, L. Vision and prey-catching strategies in jumping spiders. *American Scientist* (1982).

122. Caryl-Sue. *Bird's Eye View*, https://www.nationalgeographic.org/media/birds-eye-view-wbt/ (2015).

123. Gutierrez-Ibanez, C. Neural correlates of sensory specializations in birds (2013).

124. Ruggeri, M. *et al.* Retinal structure of birds of prey revealed by ultra-high resolution spectral-domain optical coherence tomography. *Investigative Ophthalmology & Visual Science* **51**, 5789–5795 (2010).

125. Sivak, J. G. Through the lens clearly: Phylogeny and development: The proctor lecture. *Investigative Ophthalmology & Visual Science* **45**, 740–747, doi:10.1167/iovs.03-0466 (2004).

126. Boström, J. E. *et al.* Ultra-rapid vision in birds. *PLOS ONE* **11**, e0151099, doi:10.1371/journal.pone.0151099 (2016).

127. Christensen, C. B., Christensen-Dalsgaard, J., Brandt, C., & Madsen, P. T. Hearing with an atympanic ear: Good vibration and poor sound-pressure detection in the royal python, Python regius. *Journal of Experimental Biology* **215**, 331–342 (2012).

128. Bekoff, M. & Jamieson, D. *Readings in Animal Cognition* (Mit Press, 1996).

129. Fenton, M. B. in *Hearing by Bats* 37–86 (Springer, 1995).

130. Fenton, M. B. & Simmons, N. B. *Bats: A World of Science and Mystery* (University of Chicago Press, 2015).

131. Wilson, W., Moss, C., & Thomas, J. Echolocation in Bats and Dolphins. Thomas, C. Moss (eds.) (p. 22) (2004).

132. Schuller, G. & Pollak, G. Disproportionate frequency representation in the inferior colliculus of Doppler-compensating greater horseshoe bats: Evidence for an acoustic fovea. *Journal of Comparative Physiology* **132**, 47–54 (1979).

133. Ryckegham, A. V. *How do Bats Echolocate and How are they Adapted to this Activity?*, https://www.scientificamerican.com/article/how-do-bats-echolocate-an/ (1998).

134. Moir, H. M., Jackson, J. C., & Windmill, J. F. Extremely high frequency sensitivity in a 'simple'ear. *Biology Letters* **9**, 20130241 (2013).

135. Ter Hofstede, H. M., Goerlitz, H. R., Montealegre-Z, F., Robert, D., & Holderied, M. W. Tympanal mechanics and neural responses in the ears of a noctuid moth. *Naturwissenschaften* **98**, 1057–1061 (2011).

136. Nummela, S. & Thewissen, J. The physics of sound in air and water. *Sensory Evolution on the Threshold: Adaptations in Secondarily Aquatic Vertebrates* 175–181 (2008).

137. Olson, D. & Rains, G. Use of a parasitic wasp as a biosensor. *Biosensors* **4**, 150–160 (2014).

138. Tyson, P. *Dogs' Dazzling Sense of Smell*, https://www.pbs.org/wgbh/nova/article/dogs-sense-of-smell/ (2012).

139. Keverne, E. B. The vomeronasal organ. *Science* **286**, 716–720 (1999).

140. D'Aniello, B., Semin, G. R., Scandurra, A., & Pinelli, C. The Vomeronasal organ: A neglected organ. *Front Neuroanat* **11**, 70–70, doi:10.3389/fnana.2017.00070 (2017).

141. Kimball, J. W. *Kimball's Biology Pages* (Kimball, John W., 1999).

142. Fields, R. D. Sex and the secret nerve. *Scientific American Mind* **18**, 20–27 (2007).

143. Ouma, B. O. *Population Performance of Black Rhinoceros (Diceros bicornis michaeli) in Six Kenyan Rhino Sanctuaries*, MS thesis, University of Kent, United Kingdom (2004).

144. Wesson, D. W. & Wilson, D. A. Smelling sounds: Olfactory–auditory sensory convergence in the olfactory tubercle. *Journal of Neuroscience* **30**, 3013–3021 (2010).

145. Tammet, D. *Born on a Blue Day: Inside the Extraordinary Mind of an Autistic Savant* (Simon and Schuster, 2007).

146. Abedinia, O., Amjady, N., & Ghasemi, A. A new metaheuristic algorithm based on shark smell optimization. *Complexity* **21**, 97–116 (2016).

147. Kiprop, J. *Which Animals Have The Strongest Sense of Smell*, https://www.worldatlas.com/articles/which-animals-have-the-strongest-sense-of-smell.html (2018).

148. Lee, S.-G., Poole, K., Linn Jr, C. E., & Vickers, N. J. Transplant antennae and host brain interact to shape odor perceptual space in male moths. *PloS one* **11**, e0147906 (2016).

149. Federation, W. O. *Obesity and Overweight*, https://www.who.int/newsroom/fact-sheets/detail/obesity-and-overweight (2018).

150. Baldwin, M. W. *et al.* Evolution of sweet taste perception in hummingbirds by transformation of the ancestral umami receptor. *Science* **345**, 929–933 (2014).

151. Liu, H.-X., Rajapaksha, P., Wang, Z., Kramer, N. E., & Marshall, B. J. An update on the sense of taste in chickens: A better developed system than previously appreciated. *Journal of Nutrition & Food Sciences* **8** (2018).

152. Jiang, P. *et al.* Major taste loss in carnivorous mammals. *Proceedings of the National Academy of Sciences* **109**, 4956–4961 (2012).

153. Dehnhardt, G., Mauck, B., & Bleckmann, H. Seal whiskers detect water movements. *Nature* **394**, 235 (1998).

154. Akpan, N. *Whales Can't Taste Anything But Salt*, https://www.sciencemag.org/news/2014/05/whales-cant-taste-anything-salt (2014).

155. Atema, J. Chemical senses, chemical signals and feeding behavior in fishes. In *Fish Behaviour and its Use in the Capture and Culture of Fishes* (pp. 57–101) (1980).

156. Sherman, P. W., Jarvis, J. U., & Alexander, R. D. *The Biology of the Naked Mole-rat* (Princeton University Press, 2017).
157. Barth, F. A spider's tactile hairs. *Scholarpedia* **10**, 7267, doi:10.4249/scholarpedia.7267 (2015).
158. Ahmad, H., Sehgal, S., Mishra, A., & Gupta, R. Mimosa pudica L.(Laajvanti): An overview. *Pharmacognosy reviews* **6**, 115 (2012).
159. Nicholls, J. G. & Baylor, D. A. Specific modalities and receptive fields of sensory neurons in CNS of the leech. *Journal of Neurophysiology* **31**, 740–756 (1968).
160. Kristan Jr, W. B., Calabrese, R. L., & Friesen, W. O. Neuronal control of leech behavior. *Progress in Neurobiology* **76**, 279–327 (2005).
161. Ferrier, D. E. Evolutionary crossroads in developmental biology: Annelids. *Development* **139**, 2643–2653 (2012).
162. Fijn, R. C., Hiemstra, D., Phillips, R. A., & Winden, J. v. d. Arctic Terns *Sterna paradisaea* from the Netherlands migrate record distances across three oceans to Wilkes Land, East Antarctica. *Ardea* **101**, 3–12, 10 (2013).
163. Wei-Haas, M. In *National Geographic* (2019).
164. Wiltschko, R. & Wiltschko, W. Magnetoreception in birds. *Journal of The Royal Society Interface* **16**, 20190295, doi:doi:10.1098/rsif.2019.0295 (2019).
165. Khorevin, V. The lagena (the third otolith endorgan in vertebrates). *Neurophysiology* **40**, 142–159 (2008).
166. Begall, S., Červený, J., Neef, J., Vojtěch, O., & Burda, H. Magnetic alignment in grazing and resting cattle and deer. *Proceedings of the National Academy of Sciences* **105**, 13451–13455 (2008).
167. Foley, L. E., Gegear, R. J., & Reppert, S. M. Human cryptochrome exhibits light-dependent magnetosensitivity. *Nature Communications* **2**, 356, doi:10.1038/ncomms1364 (2011).
168. Chae, K.-S., Oh, I.-T., Lee, S.-H., & Kim, S.-C. Blue light-dependent human magnetoreception in geomagnetic food orientation. *PloS one* **14**, e0211826 (2019).
169. Chalmers, J. Atmospheric Electricity Pergamon Press. *New York*, 128 (1967).
170. Morley, E. L. & Robert, D. Electric fields elicit ballooning in spiders. *Current Biology* **28**, 2324–2330. e2322 (2018).
171. Sutton, G. P., Clarke, D., Morley, E. L., & Robert, D. Mechanosensory hairs in bumblebees (Bombus terrestris) detect weak electric fields. *Proceedings of the National Academy of Sciences* **113**, 7261–7265 (2016).

172. Nuwer, R. *Respect: Sharks are Older than Trees*, https://www.smithsonianmag.com/smart-news/respect-sharks-are-older-than-trees-3818/ (2012).

173. Long, B. K. J. *How Sharks and Other Animals Evolved Electroreception to Find their Prey*, https://phys.org/news/2018-02-sharks-animals-evolved-electroreception-theirprey.html (2018).

174. Zug, G. R. *Lateral Line System*, https://www.britannica.com/science/lateral-line-system (2018).

175. Shipman, J., Wilson, J. D., & Higgins, C. A. *An Introduction to Physical Science* (Nelson Education, 2012).

176. Shashar, N. & Cronin, T. W. Polarization contrast vision in Octopus. *The Journal of Experimental Biology* **199**, 999 (1996).

177. Tichy, H. & Kallina, W. The evaporative function of cockroach hygro-receptors. *PLoS One* **8**, e53998 (2013).

178. Enjin, A. *et al.* Humidity sensing in Drosophila. *Current Biology* **26**, 1352–1358 (2016).

179. Tichy, H. & Kallina, W. Insect hygroreceptor responses to continuous changes in humidity and air pressure. *Journal of Neurophysiology* **103**, 3274–3286 (2010).

180. Ratini, M. *Does Weather Affect Joint Pain?*, https://www.webmd.com/pain-management/weather-and-joint-pain#2 (2018).

181. Gee, H. Beetles sniff out forest fires. *Nature*, doi:10.1038/news990325-2 (1999).

182. Hinz, M., Klein, A., Schmitz, A., & Schmitz, H. The impact of infrared radiation in flight control in the Australian "firebeetle" Merimna atrata. *PloS one* **13**, e0192865 (2018).

183. Stavn, R. H. The application of the dorsal light reaction for orientation in water currents by Daphnia magna Straus. *Zeitschrift für vergleichende Physiologie* **70**, 349–362 (1970).

184. Brodsky, M. C. Dissociated vertical divergence: Perceptual correlates of the human dorsal light eeflex. *Archives of Ophthalmology* **120**, 1174–1178, doi:10.1001/archopht.120.9.1174 (2002).

185. Marples, D. R. The decade of despair. *Bulletin of the Atomic Scientists* **52**, 20–31, doi:10.1080/00963402.1996.11456623 (1996).

186. Castelvecchi, D. *Dark Power: Pigment Seems to Put Radiation to Good Use*, https://web.archive.org/web/20080424001002/http://www.sciencenews.org/articles/20070526/fob5.asp (2007).

187. Dadachova, E. *et al.* Ionizing radiation changes the electronic properties of melanin and enhances the growth of melanized fungi. *PloS one* **2**, e457 (2007).
188. Balter, M. Zapped by radiation, fungi flourish. *Science NOW* (2007).
189. Dadachova, E., Bryan, R., & Casadevall, A. (Google Patents, 2014).
190. Dinc, H. I. & Smith, J. C. Role of the olfactory bulbs in the detection of ionizing radiation by the rat. *Physiology & Behavior* **1**, 139–IN137 (1966).
191. Sloan, D., Batista, R. A., & Loeb, A. The resilience of life to astrophysical events. *Scientific Reports* **7**, 5419 (2017).
192. Petrescu-Mag, I. V. Tardigrades traveled to space and survived. *Extreme Life, Biospeology and Astrobiology* **8**, 64–66 (2016).
193. Miller, W. R. *Tardigrades: Bears of the Moss (About Tardigrades)*, http://www.pathfinderscience.net/tardigrades/cbackground.cfm (1997).
194. Zantke, J., Wolff, C., & Scholtz, G. Three-dimensional reconstruction of the central nervous system of Macrobiotus hufelandi (Eutardigrada, Parachela): Implications for the phylogenetic position of Tardigrada. *Zoomorphology* **127**, 21–36 (2008).
195. Gross, V. *et al.* Miniaturization of tardigrades (water bears): Morphological and genomic perspectives. *Arthropod Structure & Development* **48**, 12–19 (2019).
196. Greven, H. Comments on the eyes of tardigrades. *Arthropod Structure & Development* **36**, 401–407 (2007).
197. Caplin, N. Anything to declare? *Physics World* **32**, 84 (2019).
198. Rothschild, L. J. & Mancinelli, R. L. Life in extreme environments. *Nature* **409**, 1092 (2001).
199. Schill, R. O., Huhn, F., & Köhler, H.-R. The first record of tardigrades (Tardigrada) from the Sinai Peninsula, Egypt. *Zoology in the Middle East* **42**, 83–88 (2007).
200. Wilcox, C. *Human-caused Extinctions have Set Mammals Back Millions of Years*, https://www.nationalgeographic.com/animals/2018/10/millions-of-years-mammal-evolution-lost-news/ (2018).
201. Khodavirdipour, A., Mehregan, M., Rajabi, A., & Shiri, Y. Microscopy and its application in microbiology and medicine from light to quantum microscopy: A mini review. *Avicenna Journal of Clinical Microbiology and Infection* **6**, 133–137 (2019).
202. Gensel, L. The medical world of Benjamin Franklin. *Journal of the Royal Society of Medicine* **98**, 534–538 (2005).

203. Vitale, S., Sperduto, R. D., & Ferris, F. L. Increased prevalence of myopia in the United States between 1971–1972 and 1999–2004. *Archives of Ophthalmology* **127**, 1632–1639 (2009).

204. Morgan, I. G., Ohno-Matsui, K., & Saw, S.-M. Myopia. *The Lancet* **379**, 1739–1748 (2012).

205. Britt, R. R. *Something Is Causing Our Eyeballs to Elongate*, https://elemental.medium.com/something-is-causing-our-eyeballs-to-elongate-df3e5dc5e371 (2020).

206. Fan, Q. *et al*. Meta-analysis of gene–environment-wide association scans accounting for education level identifies additional loci for refractive error. *Nature Communications* **7**, 1–12 (2016).

207. Hagler, L. & Goiner, L. Noise pollution: A modern plague. *Southern Medical Journal* **100**, 287–294 (2007).

208. Menkiti, N. U. & Agunwamba, J. C. Assessment of noise pollution from electricity generators in a high-density residential area. *African Journal of Science, Technology, Innovation and Development* **7**, 306–312 (2015).

209. Münzel, T. *et al*. Environmental noise and the cardiovascular system. *Journal of the American College of Cardiology* **71**, 688–697 (2018).

210. Paul, K. C., Haan, M., Mayeda, E. R., & Ritz, B. R. Ambient air pollution, noise, and late-life cognitive decline and dementia risk. *Annual Review of Public Health* **40**, 203–220 (2019).

211. Hoover, K. C. *et al*. Global survey of variation in a human olfactory receptor gene reveals signatures of non-neutral evolution. *Chemical Senses* **40**, 481–488 (2015).

212. Meddings, A. *10 Weird Food Delicacies from History that are Not Appetizing*, https://historycollection.com/10-weird-food-delicacies-from-history-that-are-not-appetizing/ (2018).

213. Yang, Q., Williamson, A.-M., Hasted, A., & Hort, J. Exploring the relationships between taste phenotypes, genotypes, ethnicity, gender and taste perception using Chi-square and regression tree analysis. *Food Quality and Preference* **83**, 103928 (2020).

214. Silk, J. B. & Boyd, R. In *Mind the gap* (pp. 223–244) (Springer, 2010).

215. Keltner, D. *Hands On Research: The Science of Touch*, https://greatergood.berkeley.edu/article/item/hands_on_research (2010).

216. Dorey, F. *How have we Changed since Our Species First Appeared?*, https://australian.museum/learn/science/human-evolution/how-have-we-changed-since-our-species-first-appeared/ (2021).

217. Jönsson, E. H. *et al.* The relation between human hair follicle density and touch perception. *Scientific Reports* **7**, 1–10 (2017).

218. De Kovel, C. G., Carrión-Castillo, A., & Francks, C. A large-scale population study of early life factors influencing left-handedness. *Scientific Reports* **9**, 1–11 (2019).

219. Balter, M. The origins of handedness. *The Science of Origins. American Association for the Advancement of Science.* Posted August 18, (2009).

220. Newland, G. A. Differences between left-and right-handers on a measure of creativity. *Perceptual and Motor Skills* **53**, 787–792 (1981).

221. Hardyck, C., Petrinovich, L. F., & Goldman, R. D. Left-handedness and cognitive deficit. *Cortex* **12**, 266–279 (1976).

222. Guy-Evans, O. *Lateralization of Brain Function*, https://www.simply psychology.org/brain-lateralization.html (2021).

223. Sweet, R. C. Prosthetic Body Parts in Literature and Culture, 1832 to 1908. (2016).

224. Luo, Y. H.-L. & Da Cruz, L. The Argus® II retinal prosthesis system. *Progress in Retinal and Eye Research* **50**, 89–107 (2016).

225. Wang, V. & Kuriyan, A. E. Optoelectronic devices for vision restoration. *Current Ophthalmology Reports* **8**, 69–77 (2020).

226. National Research Council. Hearing loss: Determining eligibility for social security benefits (Washington DC: National Academies Press, 2004).

227. Otto, S. R., Brackmann, D. E., Hitselberger, W. E., Shannon, R. V., & Kuchta, J. Multichannel auditory brainstem implant: Update on performance in 61 patients. *Journal of Neurosurgery* **96**, 1063–1071 (2002).

228. Weintraub, K. *Building a Brain Implant for Smell*, https://www.scientificamerican.com/article/building-a-brain-implant-for-smell/ (2019).

229. Lee, J. S. *et al.* Bio-artificial tongue with tongue extracellular matrix and primary taste cells. *Biomaterials* **151**, 24–37 (2018).

230. Pylatiuk, C., Kargov, A., & Schulz, S. Design and evaluation of a low-cost force feedback system for myoelectric prosthetic hands. *JPO: Journal of Prosthetics and Orthotics* **18**, 57–61 (2006).

231. Logan, D. W. (The Company of Biologists Ltd, 2021).

232. Ortiz-Catalan, M., Mastinu, E., Sassu, P., Aszmann, O., & Brånemark, R. Self-contained neuromusculoskeletal arm prostheses. *New England Journal of Medicine* **382**, 1732–1738 (2020).

233. Chandrasekaran, S. *et al.* Evoking highly focal percepts in the finger-tips through targeted stimulation of sulcal regions of the brain for sensory restoration. *Brain Stimulation* **14**, 1184–1196 (2021).

234. Serruya, M. D., Hatsopoulos, N. G., Paninski, L., Fellows, M. R., & Donoghue, J. P. Instant neural control of a movement signal. *Nature* **416**, 141–142 (2002).

235. Musk, E. An integrated brain-machine interface platform with thou-sands of channels. *Journal of Medical Internet Research* **21**, e16194 (2019).

236. Liao, X. *et al.* A bioinspired analogous nerve towards artificial intelli-gence. *Nature Communications* **11**, 1–9 (2020).

237. Gordon, R. *Teaching Artificial Intelligence to Connect Senses like Vision and Touch*, https://news.mit.edu/2019/teaching-ai-to-connect-senses-vision-touch-0617 (2019).

238. Bennett, D., Bennett, D. J., & Hill, C. *Sensory Integration and the Unity of Consciousness* (MIT Press, 2014).

239. Koch, C., Massimini, M., Boly, M., & Tononi, G. Neural correlates of consciousness: Progress and problems. *Nature Reviews Neuroscience* **17**, 307–321 (2016).

240. Poole, D., Mackworth, A., & Goebel, R. Computational Intelligence (1998).

241. Fjelland, R. Why general artificial intelligence will not be realized. *Humanities and Social Sciences Communications* **7**, 1–9 (2020).

242. Dilmegani, C. *When will Singularity Happen? 995 Experts' Opinions on AGI*, https://research.aimultiple.com/artificial-general-intelligence-singularity-timing/ (2017).

243. Stapp, H. P. Quantum approaches to consciousness (2006).

Index

CPSIA information can be obtained
at www.ICGtesting.com
Printed in the USA
JSHW050355250622
27351JS00001B/3

9 789811 246296